Basic Approaches
to Ecology

by

A. V. Shepley, M.Sc.

Lecturer in Environmental Science,
Northumberland College of Education

Pergamon Press

OXFORD · LONDON · EDINBURGH · NEW YORK
TORONTO · SYDNEY · PARIS · BRAUNSCHWEIG

Pergamon Press Ltd, Headington Hill Hall, Oxford, 4 & 5 Fitzroy Square, London W.1.

Pergamon Press (Scotland) Ltd., 2 & 3 Teviot Place, Edinburgh. 1.

Pergamon Press Inc., Maxwell House, Fairview Park, Elmsford, New York 10523.

Pergamon of Canada Ltd., 207 Queen's Quay West, Toronto 1.

Pergamon Press (Aust) Pty. Ltd., Rushcutters Bay, Sydney, New South Wales.

Pergamon Press S.A.R.L. 24 Rue des Écoles, Paris 5ᵉ.

Vieweg & Sohn, GmbH, Burgplatz 1, Braunschweig.

Selected Topics in Biology series edited by Mrs. H. Spiers

Basic Approaches to Ecology

Other titles in this series

The Bacteria by B. Phillip

Viruses by G. Usher

Genetics by P. Burke

Other Titles of Interest:

Chemistry for Biologists by T. Jennings

Multiple Choice Questions in Elementary Science

Biology by J. B. Cook

BASIC APPROACHES IN ECOLOGY

© A. V. Shepley 1969

First published 1970

08 006598 8

Contents

For Angela

Introduction

THIS book has been written in an attempt to provide a background to first ecology courses in sixth forms and Colleges. As such it does not include detailed discussions of apparatus and methods which are readily available elsewhere. Which of these are used will depend on the particular teacher, the situation in which he finds himself, and the aims of his course. Students need to know, however, that the approaches they are able to use and their methods of investigation are not the whole of ecological science.

Part II and the Appendix deal only with techniques of use in general plant and animal ecology. Those with special interests in micro-organisms or parasites, for example, may well emphasize these in their teaching and will, I hope, find opportunity to integrate them into their use of Parts I and II.

I should like to acknowledge the valuable criticism of Bryan Coultas, Lynden Emery, Kenneth Fenton, Jean Harris and Morag McGrath who have all read various parts of the text and to thank Jack Nicholson, who took the photographs for Plates 1, 2, and 3. Hilary Spiers and Bob Lowrie, of Pergamon Press have been most helpful throughout the preparation of the book. I wish also to thank Penguin Books Ltd. for allowing quotation of the definition of ecology from Abercrombie M. et al. 1954, *A Dictionary of Biology*, and Holt, Rinehart and Winston, for allowing quotation of the definition of ecology from p. 3 of Odum E. P. 1963, *Ecology*. Table XIII is taken from Table 8 of Lindley D. V. and Millar J. C. P., *Cambridge elementary statistical Tables*, by permission of the publishers, Cambridge University Press. I am indebted to the literary executor of the late Sir Ronald A. Fisher, F.R.S., to Dr Frank Yates F.R.S., and to Oliver and Boyd Ltd., Edinburgh for permission to reprint Tables XIV, XV, and XVI from their book *Statistical Tables for Biological, Agricultural and Medical Research.* Chicago University Press have kindly agreed to the use of material from Brower J. van Z. *Experimental studies of Mimicry* IV 1960 Amer. Nat. pp. 271–282. Any errors remaining are entirely attributable to myself. I would welcome comment, criticism, and correction of errors of fact.

A. V. Shepley,
Northumberland College.
November, 1968.

To the teacher: the scope of ecology

DEFINITIONS of ecology have been framed by many authors, but only those so broadly based as to be of little help to a beginner are likely to be generally acceptable to ecologists themselves.

"ECOLOGY: study of the relations of animals and plants, particularly of animal and plant communities, to their surroundings, both animate and inanimate."
   Abercrombie et al. (1954)
". . . the study of the structure and function of nature . . ."
   Odum (1963).

The main emphasis of ecology is on the relationship between organisms, and groups of organisms, and their external environment. The lack of distinction in the boundaries of ecology becomes obvious when considering parasites, e.g. tape-worm, where the physiology of its host is an important feature of its ecology, or in a treatment of behaviour where the behaviour of black-headed gulls in a colony towards predators must be part of any ecological view of predators and prey. Ecology is thus more of an outlook on the living world than a study of a specific area of it.

Ecology not only overlaps with other areas of biological study but also with subjects all too frequently regarded as separate from biology, particularly in schools. The work of Godwin, and others, on pollen analysis of peat borings, soils, and sediments has dovetailed archaeological and biological evidence of our post-glacial flora in a way which would have otherwise been impossible. The impact of man on post-glacial Britain was of both biological and sociological importance. The connecting links between geology, agriculture, and natural vegetation are a familiar study for geographers. That the basic viewpoint is an ecological one is a fact which too few sixth formers appreciate.

With an ecological approach to biology the door stands wide for those who wish to emphasize the involvement of man with the world around him. The personal involvement of pupils with their ecological studies presupposes that somewhere those studies should touch their own lives. Only when their own environment has been seen through ecological eyes can other environments be successfully investigated and compared. The field-centre based "crash" course in ecology has often been misplaced and mistimed in our teaching. Slowly, viewpoints are changing and a sound basis is being prepared before throwing the book-fed, building-surrounded, ecologically-blind majority of our children to the many-headed monster of oak wood, sea shore, limestone grassland or other equally unfamiliar surroundings.

As ecological research has continued, approaches to study have grown and expanded. The observational, natural history, approach is an initial step in all ecological work and other approaches—numerical, genetic, energetic—are progressions from it. Rather than being thought of as a progressive series in achievement (even assuming any two workers would agree on their order) these approaches should be thought of as the probing sense organs of ecological science. Continually moving, continually feeding back information, they uncover the latent potentiality of their parent body.

Each teacher, each student, must begin by learning to observe effectively. Only when the basic technique is mastered is it possible to explore further. To some degree it is, of course, possible to use the techniques of the growing points of ecology whilst still learning to observe. Indeed something of this process must continue for all scientists throughout their scientific lives or their effectiveness must diminish. It is, however, all too easy to become absorbed in the mastery of technique and to lose sight of its purpose. The aim of this book is to help teachers and their students to avoid this.

To the student: the beginnings of ecology

THE word ECOLOGY is barely 100 years old, having been coined in 1869 by the German biologist Ernst Haeckel (from the Greek *oikos*—a house, or place to live, and *logos*—a discourse). As a separate field of biology, ecology became established only in the early years of the twentieth century and has only passed from research workers to teaching in the last thirty years or so.

Although compounded of Greek roots the Greeks had no word for "ecology". Their concept of the living world was based on a small amount of observed fact, and a great deal of hearsay. Aristotle (384–322 B.C.) and his pupil Theophrastus (372–287 B.C.) both made attempts at classifying living organisms. In both cases it is noticeable that the divisions they adopted are more suited to an ecological point of view than a taxonomic one. Theophrastus, for example, distinguished a) plants living in sea water, b) plants of the sea shore, c) submerged fresh water plants, d) plants of shallow lake shores, e) plants of stream sides, and f) of marshes. 300 years later Pliny the Elder (23–79 A.D.) wrote, "A soil that is adorned by tall and graceful trees is not always a favourable one except of course for those trees. What tree is taller than the fir? Yet what other plant could exist in the same spot?" Thus he illustrated plant association, or rather the lack of it in this case.

For 1,000 years, and more, ecology together with the greater proportion of the rest of science remained static, or even perhaps retrogressed as some of the observations of classical writers were passed by. Only with the Rennaissance and its ferment of ideas and exploration in every field did ecology advance again. The German herbalist, Valerius Cordus (1515–1544), exemplifies the quickening interest, recording not only the description and properties of plants, but also their associated soils and rocks. Robert Boyle (1627–1691), best known for his physical and chemical work, interested himself in the effects of reduced pressure on animals. He particularly noticed the resistance of cold-blooded animals, and experimented further with new-born kittens:

"Being desirous to try, whether Animals, that had lately been accustomed to live without *any*, or without *full* Respiration, would not be more difficulty or slowly killed by the want of Air. . . ." and found that: "These tryals may deserve to be prosecuted with further ones, to be made not only with such kittens, but with other very young animals of different kinds, for what has been related it appears, that those Animals continued three times longer in the Exhausted Receiver, than other Animals of that bigness would probably have done."

The relevance of such work to the ability of organisms to function at high altitudes is apparent.

Linnaeus (1707–1778) and Buffon (1707–1788) stand out among the biologists of their time. Both are mostly remembered for their contributions to the solution of the problems of taxonomy but both, particularly Buffon, took account of the surroundings of organisms. From then on biologists and naturalists rapidly increased their knowledge and their specialization and it was inevitable that, towards the end of the nineteenth century, the interrelation of organism and environment should have its turn. Parallel with this the Industrial Revolution, at least in Western Europe and the United States, resulted in the most rapid change of man's environment since urbanization began. Biological science was not prepared to record that change and is only now reaching an organized and efficient enough state to accomplish such work.

As knowledge has accumulated so it has become apparent that a precise definition of **ecology** is difficult, if not impossible, and that once we start trying to be more detailed than saying simply, "Ecology is the study of the relationships of organisms with the world around them, both animate and inanimate", we find ourselves trying to express the whole changing kaleidoscope of life in too few words, and consequently failing to do so. This book, and its companion volumes, is meant to help you to think about, and to investigate, the relationships between organisms and their environment and, in particular, you and your environment. This, for you, may be the beginning of ecology, you may already have studied ecology before, you may never study it, as such, again. Whatever state you find yourself in, now and in the future, think about the changes in the world about you and think about those changes as they affect you. Try taking an ecological viewpoint when solving the problems presented by these changes. You may find some unexpected, practical, and satisfying answers.

Chapter I

PART 1

The observational basis of ecology

THE acquisition of scientific knowledge proceeds in three ways. Firstly, any scientific discipline, or branch of it, passes through the observational phase. Secondly, scientific method is applied to the problems uncovered by observation and lastly, but by no means least, what we can only call intuition can play a part. It is noticeable, however, that an intuitive insight into a problem only comes to those who have a sound basis of observation and scientific discipline behind them. In this chapter we shall look at what observation means in the context of ecological study, and at what information we might expect from simple observations without carrying the matter any further.

However sophisticated our techniques of dealing with our observations may be, however closely we may be working to the present frontiers of knowledge, and however much knowledge we already have, it is as well to bear in mind one thing; observations are made up of what is observed and the observer, changes taking place in space or time in the world around us, and you or I. In order to be sure that we are both observing the same thing all our preconceptions, any disability of sense we may have, and our ability to express ourselves must be taken into account. From this problem, as we shall see, arises our need to use numerical values and instruments to measure them. But what, you ask, do we do when instruments do not measure what we thought they measured; what do we do when you and I have agreed on an approach to a problem which turns out to be a useless one? In the first place, of course, we fail to solve the problem, and in the second place we must think it out again in the light of our previous negative experience.

Anyone who has walked in woodlands, or who has played on our rocky shores, or climbed on the mountains of Britain will have seen, although they may not have been conscious of it, that the Lichens are an important part of the scenery of these places. They cover trees, rocks, and soil in an interrupted patchwork of grey, green, black, red, and orange-yellow. No more than a casual glance will show that the centres of population do not possess this rich Lichen flora. But how subtle are the differences?

O. L. Gilbert has examined the Lichens in the area of Newcastle upon Tyne. At ten sites along a ten mile line running north-west from the city centre, he identified and counted the total number of species present on the boles of a dozen, well-grown, free-standing ash and sycamore trees, on several stretches of sandstone wall, and on two or three asbestos roofs. In addition sites elsewhere in the area were examined. The information obtained was gathered together to obtain a total picture of the distribution of Lichens in the lower Tyne Valley area.

Figure 1–1 shows the results of the survey along the line. If we compare the abundance of Lichens on the different substrata ten miles out of the city and in the city centre we see that the descending order of abundance, sandstone, trees, asbestos, is changed to asbestos, sandstone, trees. Why does this pattern exist? This problem too was tackled by simple observation.

In order to find out the distribution of Lichens throughout the area, rather than on one line only, groups of species, occurring together on the three substrata—ash, asbestos, and sandstone—were mapped. The results are shown in Figure 1–2.

The inner limit of the group of Lichens growing on sandstone closely parallels the outer edge of the built-up area on the north-western, windward side, but downwind it does not reappear before the coast and to the north it is fragmentary amongst the colliery towns. On trees the

1

Lichen "desert" has a similar shape, though of much larger area, and it keeps inland of most large towns to the north and south. Only asbestos sites inside the built-up area were mapped. The few that it was possible to map showed a characteristic Lichen group except in a small central area.

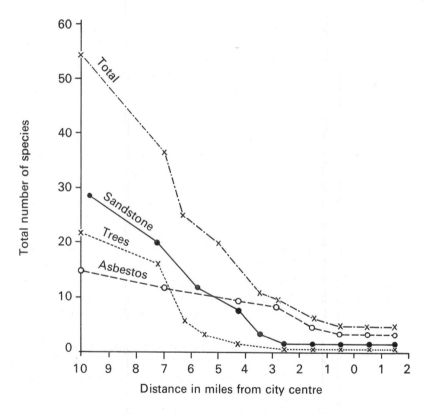

Fig. 1–1. Graph showing decreasing richness of the lichen flora as the city centre of Newcastle upon Tyne is approached. (After Gilbert O.L., *Brit.Ecol. Soc. Symposium* No. 5, 1965, p. 37).

Within Newcastle six instruments sampling the amount of smoke and sulphur dioxide in the air (as indicators of air pollution) were maintained at the time of the Lichen survey.[1] In addition two earlier series of sulphur dioxide samples were available.[2] Figures for the amount of sulphur dioxide in the atmosphere ($\mu g/m^3$) are given in Table I.

Although only one figure is available for the sandstone "desert" the general picture supports the hypothesis that the distribution of Lichens in the area is affected by the sulphur dioxide pollution in the air.

One further series of observations was made during the investigation. It was found that the acidity (as measured by pH values) of the bark of both Ash (*Fraxinus excelsior*) and Sycamore (*Acer pseudoplatanus*) decreased to within 2 km. of the city centre and in all cases was less than the average value generally recorded for these trees (Table II).

TABLE I

The sulphur dioxide content of the atmosphere of
Newcastle-upon-Tyne area

Gauge	Lichen boundary	$\mu g/m^3$		
		A.A.	W.A.	H.M.A.
N/C 17*	Asbestos	146 ± 15	175 ± 23	245 ± 22
N/C 22	"desert"	201 ± 9	296 ± 13	462 ± 35
Hexham 1	Sandstone "desert"	56 ± 0	—	—
Nafferton 1	Ash tree "desert"	40 ± 2·5	53 ± 11	75 ± 23

Adapted from O. L. Gilbert pp. 42, 43.
A.A. = Annual average, W.A. = Winter average (September to April), H.M.A. = Highest monthly average.
* The boundary of the asbestos "desert" runs between these two gauges, N/C 17 being outside the "desert".

___

[1] Filtered air is passed through hydrogen peroxide solution and the acid content determined by titration with sodium hydroxide.
[2] A lead peroxide coated cylinder is exposed to the air for a month and then analyzed for sulphates.

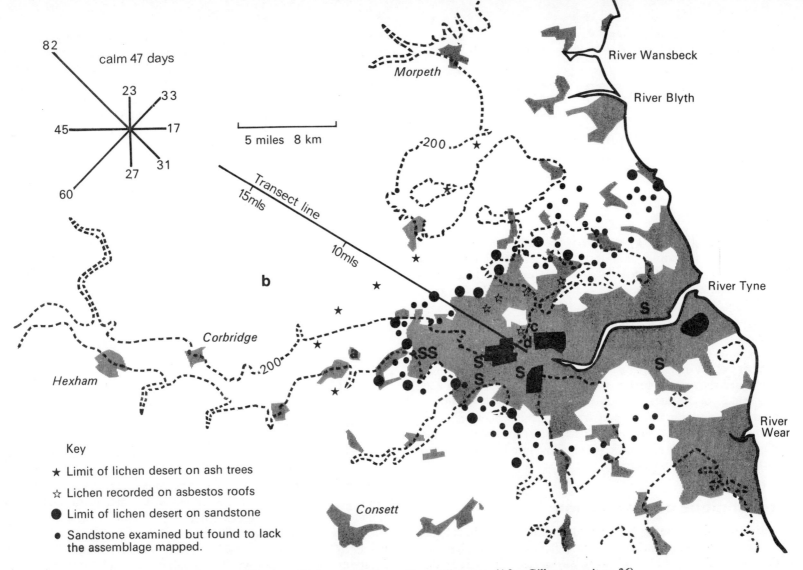

FIG. 1–2. Distribution of lichens in the lower Tyne valley area. (After Gilbert *op. cit.* p. 36).

3

TABLE II

pH of dead bark at various places to the west of Newcastle

Distance from Newcastle (km)		31	21	8·5	4	2	0
Ash	Average figure 5·5	4·2	4·5	4·0	3·8	3·4	4·2
Sycamore	6·5	4·8	4·0	4·0	3·4	3·1	3·5

Adapted from O.L. Gilbert Table 2. p. 44.

A further example

The Wood pigeon (*Columba palumbus*) is generally regarded as a pest by arable farmers. During an investigation into the relationships between numbers of Wood pigeons and their available food, R. K Murton and his associates observed the behaviour of Wood pigeons when feeding as flocks on clover. Table III records the behaviour of Wood pigeons on sighting a feeding flock.

The larger the feeding flock under observation the fewer the number of passing birds which showed no reaction and flew past, and the fewer the numbers which dipped and circled only. In the larger flocks birds which alighted showed less preference for the front or rear of the flock over the middle of the flock, though almost twice as many alighted at the front as against the middle and rear. In smaller flocks alighting birds avoided the middle of the flock. When the reactions of all alighting birds are included in an arithmetic mean, avoidance of the middle of the flock is still an obvious feature.

The technique of observation

Let us now consider the work we have described above—the study of Lichens in built-up areas, and the feeding of flocks of Wood pigeons—from the point of view of how this work was carried out. Imagine that we wished to check these observations so that we could be sure that answers could be found to questions we have in mind arising from this work. Let us assume that the only information available to us to start with is what is printed here. We must think out, from the beginning, what must be done. It doesn't matter whether we shall be repeating exactly what Gilbert, or Murton and his co-workers' did. You will see, as you gain experience of ecological work, that if the problems are to be tackled efficiently, then we must adopt an approach which is very similar to the one they must have used.

TABLE III

The behaviour of flying Wood pigeons on sighting a feeding flock.

Size of feeding flock	No. of flocks observed	No. of birds arriving in vicinity of feeding flock	% birds arriving which showed no reaction	% birds arriving which dipped or circled only	Alighted Rear	Middle	Front
1–20	4	382	35	19	22	3	21
21–100	7	282	10	2	25	4	49
101–150	3	263	0	0	29	25	46
Totals	14	927					
Arithmetic mean			18	8	28	10	36

Adapted from Murton et al. 1966, J. App. Ecol. 3:1, p. 70.

The first point of importance is that we begin with a **background** of information from our own **experience**. This background may be meagre or extensive, but without some background experience of similar nature we wouldn't even ask questions in the first place. Our background experience needs **organization**. Do we know what a Lichen is? Can we decide whether one Lichen is the same as, or different from, another? Can we identify a Wood pigeon on sight under all weather conditions? If not, what weather conditions can we not make observations under? And so on.

Organization of background material must be supported by organization of the new material as we acquire it from our observation. In other words we need accurate, organized, **written records** and accurate **identification**, or at least identification to a specified degree of accuracy. In identification, experience and practice are important. This is more obviously so in the case of Lichens than in the case of Wood pigeons, though it would at least be necessary to learn their flight pattern at a distance and perhaps against the light.

The organization of written records will involve decisions about when to use description, or when to use a map or sketch, or perhaps a photograph. Whichever method we use, our **terminology** and **symbols must be standard** to avoid confusion, at least for these investigations, and where generally accepted ones exist we should use them.

Initially, our observations will be in the form of **field notes** together with background information, gathered from various sources, which we shall have to accept as accurate either because we have no time to check it, or because we do not have the ability to do so ourselves in any case; sulphur dioxide concentrations in the atmosphere of Newcastle fall into this category.

In using these sulphur dioxide readings at all we are suggesting that the sulphur dioxide concentration in the air is a **habitat factor**, a part of the **environment**, for the Lichens we have observed. We are thus **classifying** part of our observations, and to draw any conclusion from them we shall have to do this continually as each new piece of information comes to hand.

In repeating observations on flocks of feeding Wood pigeons we shall be watching the behaviour of birds. At the same time their place, time, and method of feeding is part of the ecology of both Wood pigeons, and the clover they feed on. There is an interrelationship between one organism and another which is **dynamic** and forever changing as the populations of birds and of clover plants change. The boundaries of our ecological study must thus be arbitrary and the line where ecology stops and behaviour or, in the case of the Lichens, physiology or meteorology begins, must remain flexible to our purposes.

So far we have barely mentioned the tools of our observations. The omission has been deliberate, as different purposes require different tools and some of these will be mentioned in the other books of this series. In the main those required for the investigations we are supposing we wish to carry out are simple and easily obtained—a pencil, a notebook, a pair of binoculars, and perhaps a microscope. More complex problems might require the use of more complex tools. We must not, however, lose sight of our problem in mastering the technique of using the instrument required.

Some of the words in the preceding paragraphs have been emphasized. Many of them—organization, field-notes, standard terminology—are important throughout biology, and in other branches of human knowledge too. A few, however, deserve further consideration because they include within their orbit some of the fundamental ideas of ecology and because the recognition of their importance arose from observational ecology and they have led to both a broadening of ideas, and a more satisfactory delineation of them.

Some ecological concepts

We have already decided to leave our definition of **ecology** as "the study of relationships of organisms with the world around them, both animate and inanimate". Eventually you will have to decide for yourself whether or not this definition is an adequate one for general use. At the moment, however, it can be said that you need to know a good deal more of the concepts of ecology before you can interpret the results of your observations. Earlier workers have had to both interpret their results and build their concepts at one and the same time. In this we are at an advantage, but only by comparing our own work with theirs shall we reach the heart of the concepts which form the frame-work of ecology.

Let us look at the **environment**. This is everything outside the body of the organism. It is unlikely that anyone will succeed in studying the environment of any organism from all aspects, though some have come near to completely covering the whole of the **autecology** of a single

5

organism. More usually the scale of the investigations makes it necessary to single out various features of the environment—**habitat factors**—for special study. On the whole we shall tend to select features which we think, at least at the start of an investigation, are important to an organism, features which make up its **effective environment**.

An alternative approach to the environment is to treat it in geographical units, to divide it into a field, a wood, a log, or a reed-bed, and to study the interrelations of all organisms within that area. Such a study would be **synecology**. You will see that of our two examples, the Wood pigeon study is essentially autecological, the Lichen study essentially synecological.

The ecological study of a **habitat**, or environmental unit, will soon show that organisms behave differently when in groups from their behaviour as individuals. This is well shown by the pull of the feeding flock for the individual pigeon. The group concerned here is the **species**. Regarded by many biologists as the most fundamental group of all, the species is kept together by its pattern of breeding and exchange of genetic material. Species do not, however, have uniform **distributions**, but overlap irregularly with the distribution of other species. Such irregular overlapping results in interaction which produces the characteristic assemblages of species in habitats known as **communities**. The uneven distribution of species also results in breeding units within the species smaller than the total population which are known as **demes**.

The individual, the community, and the deme all show mechanisms which tend to maintain the *status quo*, and keep variation within narrow limits. An example is the control of body temperature by mammals. Browning has shown that such **homeostatic** behaviour operates at the deme level amongst populations of the kangaroo-tick (*Ornithodorus gurneyi*) in western and southern South Australia. Here the camping places of the red kangaroo, on which it feeds, are so abundant as to make the general incidence of *Ornithodorus* low, for kangaroos will tend to move on quickly and the chances of the tick falling off after feeding in places where it can later attach itself to another host are low. The **dispersion** of the ticks is thus in an uneven pattern as a result of interaction with their environment, both animate and inanimate. The actual movement of ticks from one place to another, or their **dispersal**, is distinct from their dispersion and is not directly connected with the suitability of an area for colonization by the species.

Dynamic ecology

In the same way that in the life-time of one man the hills and mountains he knows appear to change little, so, on the face of things, most of the concepts outlined above apply to unchanging communities of organisms. A very little knowledge of geology opens one's eyes to the erosive powers of water, ice, and wind and soon you see all around you changes in the face of the land. Similarly communities or organisms have their dynamic aspect apart, that is, from the ability of individuals to move from place to place.

Brady surveyed the fauna of the cow-pats produced by Hereford bullocks. The presence of various organisms was recorded over a period of two months and is summarized in Table IV.

The occurrence of larval stages only after the adults had already been recorded can be seen. The adults, however, appear and disappear again in a sequence. *Scatophaga stercoraria* appears within an hour of deposition of the dung and ceases to visit it within 24 hours. *Limosina* also appears within an hour or two but continues its visits for several days. By the second week there is a firm crust over the dung and Dolichopodid flies visit the softened pats after rain and the two scarabs, *Aphodius erraticus* and *A. fossor*, together with Staphylinid beetles burrow in it. Cyclorrhapha and Orthorrhapha larvae also present are most likely to have been feeding on the dung itself, while the Dolichopodid flies fed on them.

By the fourth week Scarabaeid and Staphylinid larvae are present and Braconid wasps (perhaps parasitizing the Staphylinid larvae) visit the pats. The dung has become fibrous and Collembola, *Allolobophora terrestris* (an earthworm) and fungi are now present.

Such a sequence of organisms in a habitat, or **micro-habitat**, as it may be called in this case, is known as a **succession**. This succession, beginning with arthropods, followed by lumbricidae, as the dung ages is a

TABLE IV

Qualitative summary of incidence of fauna relative to age of cow-pat

	0	2	4	6	8	0	2	4	6	8	0	2	4	6	8	0	2	4	6	8	0	2	4	6	8	0	2	4	6	8
Limosina adults	*	*	*	*																										
Scatophaga adults	*																													
Lucilia adults	*	*																												
Dolichopodidae adults					*	*																								
Scarabaeidae adults		*				*			*					*			*													
Staphylinidae adults			*	*			*	*	*		*	*		*																
Cyclorrhapha larvae					*	*	*	*	*	*	*																			
Orthorrhapha larvae					*	*	*	*																						
Scarabaeidae larvae											*	*	*	*	*															
Staphylinidae larvae											*	*	*	*	*		*		*											
Braconidae adults											*	*		*																
Collembola adults																*		*		*										
Lumbricidae																					*	*	*	*	*	*	*	*		

Age of pat in days after deposition, as consecutive periods of eight days   * Marks incidence.

From Brady C. *Cow pats as an ecological habitat* 1965 S.S.R. XLVI No. 160 p. 625

primary succession. The recording of fungal hypae at the fifth week suggests that observation might show a **secondary succession** of fungi as the breakdown of the dung continues further. Succession is by no means confined to micro-habitats or to arthropods and fungi, but is a feature observable in any habitat which becomes freshly available to invasion by organisms. It has proved possible to identify successive plant communities in many habitats the whole succession being termed a **sere** and being named after a characteristic and over-riding environmental feature e.g. a succession from fresh-water to land is known as a **hydro-sere.**

Another feature of the study of cow pats referred to above is that more than one organism may be feeding on the pats at the same time. There must be **competition** for food between them. In other situations competition for light, water, or shelter might arise between organisms. Some of the organisms feeding within the pats are feeding on the material of the pats, others feed on them. Thus a picture of the feeding relations of the cow pat organisms can be built up to produce a **food-web** of **food-chains** as shown in Figure 1–3. Notice how, though both are carnivorous, the Dolichopodid flies and *Scatophaga* are not in direct competition, but each has its ecological **niche**, the one feeding on Cyclorrhapha and Orthorrhapha larvae and the other on *Limosina*.

Can simple observation take us further?

In his classic work "Animal Ecology" Charles Elton wrote: "If you are studying the fauna of an oak wood in summer, you will find vast numbers of small herbivorous insects like aphids, a large number of spiders and carnivorous ground beetles, a fair number of small warblers, and only one or two hawks. . . . To put the matter more definitely, the

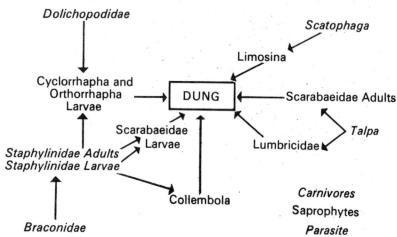

Fig. 1–3. The food web on cow pats. (After Brady C., *S.S.R.*, 1965, 46, p. 625).

animals at the base of a food-chain are relatively abundant, while those at the end are relatively few in numbers, and there is a progressive decrease in between the two extremes. . . . The general existence of this *pyramid in numbers* hardly needs proving, since it is a matter of common observation in the field. Actual figures for the relative numbers of different stages in a food-chain are very hard to obtain in the present state of our knowledge".

This last statement still applies. Suppose we wished to arrive at figures for the food-chains Braconid wasps—Staphylinid larvae—Scarabaeid larvae or, *Talpa* (mole)—Lumbriciade. Even assuming we could be content with working on one cow pat only—and this might easily be different from the majority of the remainder—we still have to catch our organisms and identify them. This would not be easy with mobile animals, and we could not be sure that we had all of them at any one time.

With problems such as this all we can do is settle for an estimate of the numbers of organisms involved. In other cases as well we may have to accept estimates, degrees of probability of being correct. It is not easy to compare different communities of organisms in time, or in space, by simple observation. At this point we must adopt another method of approach. This will be the subject of Chapter II.

Summary

1. The acquisition of ecological knowledge must be based on careful observation.
2. The organization of observation requires recording, accurate identification, and classification of information.
3. Observational techniques have produced ecological concepts which involve a dynamic relationship between the species on the one hand, both as individuals and as communities, and the environment on the other hand.
4. Eventually a point is reached where simple observation cannot give answers to certain questions which are suggested by the observations. At this point a new approach must be adopted.

Chapter II
Numerical aspects of ecology

At the end of the last chapter we reached the conclusion that a fresh approach to our observations is needed if we are to extend their usefulness. A new depth can be given to the analysis of the results of our observations by making use of the mathematical tool known as **statistics**. The mathematical theory on which statistical techniques are based is the province of specialist statisticians whose advice we can ask at more advanced stages of study. Meanwhile the detailed nature of their work is of as much importance to us as the design aspects of engines are to the amateur motor mechanic. We are forced to take the fashioning of our statistical tools on trust but we do not have to understand the finer points of statistics in order to use statistical method. The details of statistical methods useful in ecology will be found in the Appendix. In this present chapter we are concerned with exploring some of the new fields opened up in ecology by the use of a numerical approach.

Even the simplest observations are often recorded in terms of numbers. Sometimes there are obvious patterns in the numbers and we need proceed no further to find the answers to our questions. At other times there is no apparent pattern, or the patterns we find provide no answers to our questions. Perhaps we may find our questions are badly phrased if we look at them closely.

An assumption

The statistical techniques we can use to help us answer our questions depend upon the numerical values obtained being distributed in a particular way. If all the values which could exist for any measurement, say the heights of men or the weights of Robins, were collected together some values would occur more than once. The total population of all the values may be graphed to produce a frequency distribution of the population (see Appendix p. 45). The assumption we must make is that the frequency distribution curve is a bell-shaped curve (Fig. A–1). Such a curve is sometimes referred to as a normal curve and describes a normal distribution which has fixed mathematical properties that can be analyzed statistically and that are independent of the scale, the magnitude, or the units of measurement used. It is, therefore, possible to tabulate many of the properties of a normal curve and to refer to the tables whenever the normal curve describes the distribution of a population. Other sorts of distribution curve do occur, but the normal curve is the most common in biology and the easiest to make calculations for. Unless there is good reason for thinking otherwise the assumption of a normal distribution can be made, at least to begin with. This decision can be made with greater confidence by ensuring, as well as we can, that the data, the numerical observations we collect, are obtained in a suitable way.

Observing data: sampling

Obviously it is only rarely possible to obtain all the possible measurements which make up the total population of values and we can only obtain some of them. Measuring only some of the values in a population is the process of sampling. It is important to take our samples in such a way that if a population of observational values is distributed normally we do not obscure this distribution by unwittingly biasing our samples. It is equally important that, if we do take trouble over our sampling methods and our observations do not conform to a normal distribution,

B.A.E.—C

we can then consider trying other sorts of distribution curve. To do this goes beyond the scope of this book.

If possible no one sample should be affected in any way by any other, i.e. samples should be at random. Many possibilities will spring to mind which seem to ensure that samples will be taken at random, e.g. throwing a device for vegetation sampling whilst turning with eyes closed—neither this nor similar methods should be used as all continue to retain a subjective bias, usually in favour of things which are conspicuous. The only satisfactory way to randomize samples is to identify them by numbers taken from a table of random numbers (see Appendix p. 48), numbers thrown on dice, or drawn from a pack of cards.

Situations frequently arise where strict random sampling is unsatisfactory. A salt-marsh area, for example, may be uneven in its distribution of plants, or an area patterned by ridge and furrow may be under investigation. In such circumstances the appropriate method is to divide the areas of interest into equal-sized sub-divisions and to take random samples within each sub-division—stratified random sampling.

Regular sampling, the exact opposite of random sampling, is also of use. Transects, for example, consist of lines laid out across environmental gradients which are sampled at known intervals along their length. The major feature of both regular and stratified random sampling is that the degree of non-randomness is known and not left to unknown factors.

How many samples?

The number of replicate samples required in an investigation depends on the statistical tests to be used on the data collected, the accuracy required, time, cost, and availability of materials. Bearing these considerations in mind it is obvious that the larger the samples the better the estimate of the population will be. A separate decision must be made for each investigation, accuracy, time, and cost having to be finely balanced against each other.

It is preferable to over-sample than to under-sample. A subjective idea of the approximate suitable sample size may be gained from a small pilot investigation. It should be borne in mind that only towards the end of an investigation can one be sure that enough samples have been taken.

An example

Much of the heather-covered moorland in the north and west of Britain is subjected to controlled burning to produce fresh heather and grass growth for grouse rearing and sheep grazing. Burning is usually on a regular cycle in any one area and consequently provides ready made sites for the investigation of how quickly vegetation becomes re-established after burning. We might also investigate whether the proportions of the various plant species present remain constant during re-colonization.

The point-frame method (described in more detail in Chapter VI p. 39) was used to compare an old established heather moor with an adjacent area which had been burnt two years previously. Fifty frames were placed randomly within a fifty metre square well inside the boundary of each area and a record made of the number of times each pin of the frame touched heather (*Calluna vulgaris*), bent grass (*Nardus stricta*), a sedge (*Carex* sp.), and various mosses, which were not separately distinguished. The results are given in Table V.

Our question about the proportions of plant species remaining constant during re-colonization can now be dealt with if we first rephrase it. If we consider the pairs of readings for each species on the two areas we are, in fact, asking "Are the two sets of values from the same total population of values, or are they to be considered as being two sets of values from the same population?" Looking at the figures in Table V you may feel that you can say that there is a difference between the values for heather on the established moor and on the area recently burnt. Is it possible, however, to say how confident you are about this? Can you easily make an assessment of the other pairs of readings?

Table V

Results of a comparison of established moorland with an area burnt two years previously. Number of pins touching any plant—total pins per frame ten.

Established moorland				Two year old burn			
Heather (*Calluna vulgaris*)	Bentgrass (*Nardus stricta*)	Sedge (*Carex* sp.)	Mosses	Heather (*Calluna vulgaris*)	Bentgrass (*Nardus stricta*)	Sedge (*Carex* sp.)	Mosses
10	—	—	4	7	1	3	—
—	—	—	—	1	—	10	—
10	—	—	1	6	—	—	1
10	—	—	—	7	—	—	—
10	—	—	—	5	6	—	—
1	—	—	—	8	—	—	—
10	—	—	—	2	—	1	2
9	4	—	1	2	—	4	4
10	—	—	—	7	—	1	—
10	—	—	—	6	—	—	4
1	—	—	—	7	1	1	—
5	—	—	—	4	3	—	—
2	4	5	2	9	—	—	1
8	—	—	—	5	—	—	—
9	2	1	—	2	—	—	—
10	—	—	—	1	1	—	1
8	—	—	—	6	—	—	—
6	—	—	—	2	—	—	3
5	—	—	—	9	—	—	2
10	—	—	—	2	2	—	4
10	—	—	—	3	—	4	4
10	—	—	—	8	—	—	2
10	—	—	6	2	—	2	6
7	—	—	—	4	—	—	6
5	—	7	1	8	—	2	3
—	—	—	1	1	—	5	2
3	—	—	2	3	—	3	1
10	—	—	—	5	—	9	—
10	—	—	—	10	—	—	—
10	—	—	—	2	—	—	—
9	—	—	3	8	—	—	—
7	—	—	—	—	7	—	1
8	3	—	—	1	—	—	2
10	5	—	—	—	—	—	9
8	—	—	—	10	—	—	—
9	—	—	—	1	—	—	2
4	—	—	2	5	—	—	7
4	—	—	3	—	—	—	5
9	4	2	—	3	4	—	1
—	—	—	—	6	—	—	3
10	—	1	—	2	1	5	—
10	—	4	—	10	—	—	—
10	—	—	—	4	—	—	2
4	—	—	—	1	5	—	1
1	—	—	—	—	8	—	—
10	—	—	—	—	—	9	—
9	—	—	1	3	—	—	5
10	—	—	3	2	1	—	6
—	1	7	—	2	—	—	—
10	—	—	—	4	6	—	—
			—				—
Total		30					90

One population or two?

We cannot give a straightforward "yes" or "no" answer to the question of whether the two sets of values come from one population or two without measuring the whole population. This is a practical impossibility and we have already dismissed the idea by taking samples in the first place. We can, however, give an answer which is qualified by a statement of the likelihood of its being wrong. We can estimate the probability of our answer being a freak answer and hence we can assess the reliability we are prepared to place on its use.

In everyday life it is surprising how much reliance we place on quite low probabilities. If something happens as a consequence of an action as few times as three in a row we tend to be confused if, at the fourth attempt, something different happens. If the same consequence occurs as many as five or ten times we become convinced that the consequence is inevitable. Many of the decisions in our lives are based on poor evidence of this sort.

When employing statistics it is necessary to make a subjective decision of how acceptable a chance is of being wrong. In many cases a 95 per cent probability is acceptable, i.e. there is one chance in 20 of being wrong. It must always be borne in mind that greater accuracy than this might be necessary. In such a case it might be necessary to take more samples to be more accurate.

TABLE VI

The initial steps in the calculation of 't' using the data of Table V for mosses.

Line	Description of statistic	Symbol	Established Moorland (x)	Two Year old burn (y)
A	Number of observations in sample	n	50	50
B	Degrees of freedom	$n-1$	49	49
C	Mean value of observations	$\bar{x} = \Sigma \dfrac{x}{n}$	0·6	1·8
D	Sum of the squares of deviations from the mean	$\Sigma d^2 = \Sigma x^2 - \dfrac{(\Sigma x)^2}{n}$	78	252
E	Estimate of variance of population	$S^2 = \dfrac{\Sigma d^2}{n-1}$	1·59	5·14
F	Estimate of standard deviation of population	$S = \sqrt{\dfrac{\Sigma d^2}{n-1}}$	1·26	2·27

Σ means "sum of".

The calculation of 't'

We can decide if there is a significant difference between the members of a pair of value sets by calculating the statistic 't' for the sets. This may then be used to find the probability of our being wrong from the appropriate table (see Appendix p. 105). The theoretical background to this and other calculations is summarized in the Appendix. For the present we are concerned only to establish that our objective can be achieved in this way.

In Table VI the initial steps in computation are set out in order for the values obtained for mosses. Some of the names of the terms used in this table and on the following pages will perhaps be unfamiliar. This will not affect your ability to follow the calculations or to do them for yourself. If you wish to understand them more fully before continuing turn to the Appendix and Glossary for their definitions.

The calculations necessary can be done quite quickly on scrap paper, especially if the individual calculations are shared out amongst a group. Calculating machines, whether hand or electric, reduce the time taken, and consequently reduce the boredom which can result from such repetitive work.

The first step in the final stage of the calculation is to set up the null hypothesis that there is no difference between the two sets of values i.e. the difference between their means is zero. Line C of Table VI shows that there is a difference between their means of $1\cdot 8 - 0\cdot 6 = 1\cdot 2$. Is this difference significantly greater than zero?

The means of both sample populations (from the old established moorland, and the area burnt two years previously) are subject to some error, which is shown by the values of their standard deviations, S (Line F Table VI). The difference between the two means will, therefore, be subject to a source of error from each of the means. The value of this error is calculated by adding together their variances, S^2 (Line E Table VI). We can now calculate the standard deviation of the difference of means (S_d) as follows

$$S_d = \frac{S^2\text{established moorland}}{n\text{established moorland}} + \frac{S^2\text{two year old burn}}{n\text{two year old burn}}$$

$$= \frac{1\cdot 59}{50} + \frac{5\cdot 14}{50} = 0\cdot 135$$

The final step in the calculation is to substitute the appropriate values in the equation to calculate 't'

$$\text{'t'} = \frac{\text{deviation of difference of means from zero}}{\text{standard deviation of difference of means}}$$

$$= \frac{1\cdot 2}{0\cdot 135} = 8\cdot 89$$

The table of distribution of 't' (Appendix p. 49) gives the maximum values of 't' shown by a normal population. If we enter this on the left hand column at the point nearest the appropriate numbers of degrees of freedom ("established moorland $^{-1}$ $^+$ " two year old burn $^{-1}$ $^-$ ^{98}) i.e. 120, we see that the difference between the means might be expected to result in a value of 't' not more than 3·29 times the standard deviation of the difference of means in 0·1 per cent of cases.

In summary then, we can accept that the chance of the sample populations being from the same total population is less than one in 1,000, and they may be regarded as being significantly different.

Other useful statistics

Several other statistical tests which are useful in ecological work are mentioned in Part III. Each of these is briefly described, and the method for calculation set out, in the Appendix.

Statistics and experiment

Ecological studies are beginning to make use of controlled experiments for testing hypotheses based upon observation. It is important to remember that controls need to be carefully selected and that some thought must be given to the design of the experiments in relation to the techniques of statistical analysis to be used on the results. In all fields of biology ill-designed experiments abound, the results of which cannot be statistically treated.

Summary

1. Where observations have been recorded numerically it is possible to use statistical analysis to derive information which simple observation cannot give.
2. If statistical analysis is to be used, then it is important that sampling of populations should be random or, in some cases, non-random in a known way.
3. The number of samples to be taken is dependent on several factors but has a minimum level below which the desired accuracy cannot be attained.
4. An example of an investigation of a moorland area subject to periodic burning is used to illustrate the calculation of the statistic 't', leading to a definite statement of the difference between the vegetation of old established moorland and a two year old burnt area.
5. Other statistical tests will be mentioned in Part III and described in the Appendix.
6. Experiments must be designed to allow specific statistical techniques to be used and not carried out in the hope that they may be statistically analysed.

Chapter III
Ecological energetics

Pyramids of number and of biomass

At the end of Chapter I we discussed the food web associated with cow pats and referred to the pyramid of numbers (Fig. 3–1, a, b). Plant material lies at the base of all such pyramids, for plants are the **primary producers** trapping radiant energy from the sun and converting it into the chemical energy of the compounds of which plants are built. In the food chain

plant $\longrightarrow$ aphid $\longrightarrow$ insectivorous bird $\longrightarrow$ hawk

we have a primary producer followed by primary, secondary, and tertiary consumers. **Primary consumers** must always be herbivores, while further consumers are carnivorous. Detritus, either partially utilized by consumers, or the remains of producers and consumers, is eaten by saprophytes and saprozoites which may be herbivorous, carnivorous or omnivorous and which are perhaps best separated simply as **detritus feeders**. Organisms feeding on other living organisms —**parasites**—show an inverted pyramid of numbers (Fig. 3–1 c). Numbers of plant parasites, for example, are larger than the numbers of plants on which they feed, and the numbers of their own parasites are, in turn, greater still.

Pyramids of number have the obvious disadvantage of equating an individual aphid with an individual hawk, a single worm with the bird which feeds on it. In addition the individual organism of a species in any one investigation is equated with an individual of the same species in any other investigation even though the one may be a juvenile and the other an adult. These difficulties can be partially removed by comparing the weights of the organisms involved (biomass) instead of their

numbers. The resulting **pyramid of biomass** (Fig. 3–1, d, e) thus represents the weight of primary producer which can support a given weight of primary consumer, and so on.

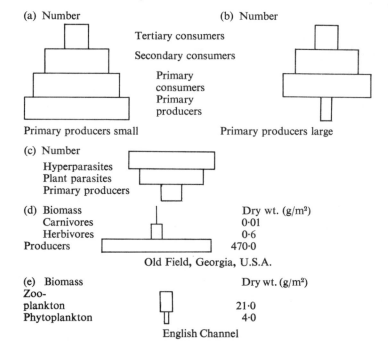

FIG. 3–1. Pyramids of number and biomass.

(a) Number — Tertiary consumers / Secondary consumers / Primary consumers / Primary producers — Primary producers small

(b) Number — Primary producers large

(c) Number — Hyperparasites / Plant parasites / Primary producers

(d) Biomass — Dry wt. (g/m²)
Carnivores — 0·01
Herbivores — 0·6
Producers — 470·0

Old Field, Georgia, U.S.A.

(e) Biomass — Dry wt. (g/m²)
Zoo-plankton — 21·0
Phytoplankton — 4·0

English Channel

Figs. 3–1 (d) and (e) from Odum E. P. 1959 Fundamentals of Ecology 2nd ed. p. 63.

A glance at Fig. 3–1 e shows, however, that this approach can produce apparently paradoxical results. Here the mass of zooplankton in the English Channel, at the sampling time, is greater than the phyto-plankton present. The key to the paradox lies in the words "at the sampling time", for change through time has not been taken into account. Zooplankton generally lives longer than phytoplankton and several generations of phytoplankton will grow and die while one of zooplankton is alive. It is, therefore, the rate at which material is transferred from one trophic or feeding level to the next which is important.

Energy in living organisms

Energy exists in various forms, radiant, electrical, atomic, and so on, all of which are interconvertible. The conversion from one sort of energy to another is accompanied by one particular energy form, heat. All forms of energy except heat, are organized, or non-random move-ment of molecules, while heat is the result of random movement of molecules. These relationships are summarized in the Laws of Thermo-dynamics—
1.   Energy may be transformed from one form to another but is neither created nor destroyed.
2.   Processes involving the transformation of energy will not take place spontaneously unless there is a degradation of energy from a non-random to a random form.

Since all other forms of energy can only be completely converted into heat it is convenient to use, as a general unit for measurement, a unit of heat measurement, and the kilogramme calorie (Kcal. or Cal.) is used. One calorie is the amount of heat required to raise one gramme of water $1°C$ from $14.5°C$ to $15.5°C$ and one Kcal is 1,000 calories. Using the appropriate factors the units used for other forms of energy can be converted into their heat equivalents. The chemical energy of substances, including living organisms, can be measured in Kcals by completely combusting them in a bomb-calorimeter.

Whether a chemical reaction takes place in one step or in several steps, since energy can neither be created nor destroyed, it follows that the total amount of heat used or evolved will be the same. Direct oxidation of hexose sugar

$$C_6H_{12}O_6 \ + \ 6O_2 \longrightarrow 6H_2O \ + \ 6CO_2 \ + \ 673 \ Kcal$$

releases the same amount of energy as the two-stage reaction

1. $C_6H_{12}O_6 \longrightarrow 2C_2H_5OH \ + \ 2CO_2 \ + \ \quad 18 \ Kcal$
2. $2C_2H_5OH \ + \ 6O_2 \longrightarrow 6H_2O \ + \ 4CO_2 \ + 655 \ Kcal$

$$\text{Total} \quad 673 \ Kcal$$

The ultimate source of energy for living organisms is the radiation of the Sun. The amount of this which reaches the Earth's surface varies from place to place. In Britain it is about $2.5 \times 10^5 \ Kcal/m^2/yr$. A lot of this is scattered by dust particles before it reaches the primary producers, even more is directly converted to heat, and more still is taken up in the evaporation of water. Probably only one to five per cent remains for transformation into chemical energy in plant tissues by photosynthesis.

Pyramids of energy

By measuring the energy contained in the various trophic levels, over defined areas of the earth's surface, and over comparable time periods, it is possible to overcome the difficulties raised by the use of pyramids of number and of biomass. Before considering an example of a pyramid of energy we must remember that, in order to carry out synthesis of organic materials at each trophic level, a certain amount of energy is required which cannot be passed on to the next trophic level but which is removed from the food chain as heat of respiration.

Figure 3–2 illustrates the broad outlines of energy changes within and between trophic levels for Silver Springs, Florida, U.S.A. The relatively high values for respiration energy coupled with the amounts retained within the trophic levels reduce the potential food energy for the next trophic level to a small proportion of the energy at any one level.

Using the measurement of energy contents of organisms it now becomes possible to directly compare communities which differ in space and time. Comparisons become valid comparisons of like with like despite the variety of community structure.

The flow of energy

Energy is constantly entering and leaving communities of living organisms. Figure 3–2 makes it possible to derive the general model for the flow of energy in communities shown in Fig. 3–3.

Using this model it also becomes possible to calculate the rate of change in the energy content of the standing crop (which may be given the general designation Λn), i.e. the rate at which energy is absorbed by the standing crop (λn) minus the rate at which energy is lost from it ($\lambda n'$).

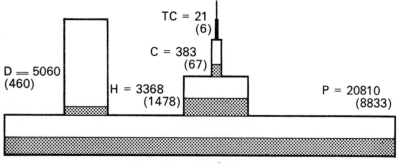

FIG. 3–2. Pyramid of energy for Silver Springs, Florida, U.S.A. (After Odum E.P., 1959, *Fundamentals of Ecology*, p. 65). The proportion of the total energy flow which is actually fixed as organic biomass, and which is potentially available as food for other populations in the next trophic level, is indicated by figures in brackets and by the shaded area at each trophic level. Trophic levels are: P = producers, H = herbivores, D = decomposers, T,TC = (Top) Carnivores.

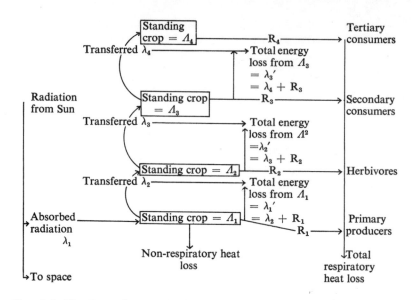

FIG. 3–3. The flow of energy in living communities. (Based on Lindeman, 1942, *Ecology* 23 pp. 399–418).

Ecological efficiency

Since, with each transformation, some energy is dispersed as heat, no utilization of a potential energy source by living organisms can be 100 per cent efficient. Measuring the relative efficiency of transformations within ecosystems provides both an indication of their basic functional mechanisms and a guide to man's needs in utilizing their production potential. Are energy relations within communities constants—does $\lambda n/\lambda n-1$ give a constant value in all communities—or are the relations different for different communities? Answers to these questions are not easy to get. Communities are usually extremely complex and the basic observation of the food web usually lacking. A beginning has proved possible using simple communities artificially created in the laboratory and some of the simpler communities found in the field.

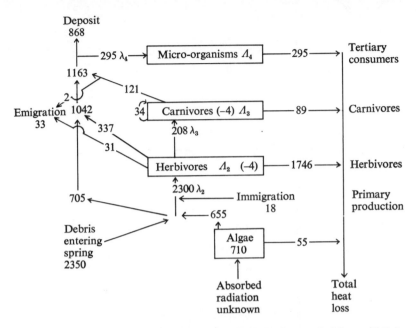

Fig. 3–4. Energy flow diagram for Root Spring, Concord, Mass., U.S.A. (Re-drawn from Teal J. M., 1957, *Ecol. Monogr.* 27, pp. 283–302).

The comparison of $\lambda n/\lambda n{-}1$ in Table VII shows that the value for $\lambda 4/\lambda 3$ for Root Spring deviates widely from the others. In this case $\lambda 4$ is built up of increments from all previous trophic levels, and the tertiary consumers (micro-organisms) are detritus feeders. A maximum value of 12·5 per cent—quite close to the other field values given above—was obtained in a laboratory experimental food chain by Slobodkin. This consisted only of a primary producer—the alga *Chlamydomonas*; a primary consumer—the crustacean *Daphnia*, and man acting as a secondary consumer by removing individual *Daphnia*.

Energy change in ecosystems

Study of the relatively simple communities already mentioned illustrates the difficulties of applying an energetic approach to whole ecosystems. At the moment we are too ignorant of the details of action and interaction of organisms in such complex ecosystems as woodlands or large lakes, for example, to enable us to follow through the energy changes entailed in feeding, growth, reproduction and respiration in such ecosystems. It is possible, however, to tackle some parts of the problem, especially photosynthetic and respiratory activities.

Figure 3–4 shows the energy flow diagram determined by Teal for Root Spring, Concord, Mass., U.S.A. during 1953–54. This small spring received energy from both primary producers and debris entering it. In no case has the standing crop energy been calculated. The standing crop at both Λ_2 (herbivores) and Λ_3 (carnivores) was slightly lower (-4 Kcal/m²) at the end of the year than it was at the beginning.

TABLE VII

Comparison of the ecological efficiencies of Root Spring and Silver Springs

	λ3/λ2	λ4/λ3
Root Spring	8·97%	114·18%
Silver Springs	9·75%	16·65%

Summary

1. Pyramids of number and of biomass are unsatisfactory when comparing different ecosystems. Difficulties are removed by considering energy changes within ecosystems.
2. Using an appropriate model it is possible to diagram the flow of energy through an ecosystem.
3. The relative efficiencies of various ecosystems can be compared and are potentially particularly relevant to the understanding of the management of environment by man.
4. Information available so far has enabled us to do no more than make a start in understanding environmental management problems.

Chapter IV

Ecological genetics

1ST JULY, 1858 represents a land mark in the history of biology, and indeed in the history of mankind. On that day Charles Darwin and Alfred Russell Wallace presented a joint paper to the Linnean Society of London. This was the first public statement of what we now know as the Theory of Evolution.

Evidence in favour of this hypothesis is regarded as overwhelming by most biologists. This has not been consistently so since 1858, nor has evidence accumulated in a continuous even flow. Following the re-discovery of the results of Gregor Mendel's genetic studies the early 1900s saw the development of a strong trend against Darwinism. This was still loudly voiced during the 1930's. A reversal of this trend only became possible (a) with the full appreciation brought about by the study in depth of the genetics of a few laboratory bred organisms and (b) with the gathering of field evidence that evolution can be shown to be going on all the time under natural conditions often at much faster rates than Darwin had thought possible.

The development of a new field of study, in this case the study of evolution in action, often requires the development of new techniques, but even more important it requires a fresh mental approach. That this has been successfully done, integrating in a new way scientific observation and experiment in both field and laboratory, is largely due to the work of E. B. Ford during the last half century. He has called this combined area of study Ecological Genetics. In this chapter we shall look at some aspects of Ecological Genetics.

Polymorphism

Though polymorphism literally means "many formed" we are concerned here only with the occurrence together in the same habitat, at the same time, of two or more distinct forms of a species in such proportions that the rarest of them cannot be maintained merely by recurrent mutation of genes. This excludes some familiar ways in which a species may be many formed e.g. geographical races, and seasonal forms.

Industrial melanism

In Chapter 1 we described the distribution of lichens in the Newcastle upon Tyne area and its relation to the pollution of the atmosphere by industry and the burning of domestic fuels. Atmospheric pollution has also resulted in changes in the populations of moths in industrialized areas known as **Industrial melanism**. This phenomenon has been observed on the continent of Europe, and in the United States also, while in Britain over eighty moth species have been affected. In this instance, at least, biologists have successfully recorded the after effects of the Industrial Revolution.

Perhaps the most well-documented case of industrial melanism is that of the Peppered moth, *Biston betularia* (see Plate 4). Before about 1850 only the typical black and white speckled form of this moth had been caught by entomologists. Since that time two darker forms of the moth have become prevalent in industrial areas (Fig. 4–1).

a) *B. betularia carbonaria* (Plate 4)

b) *B. betularia insularia*.

Carbonaria is almost entirely black. It was first noticed in the Manchester area in 1848 but has now spread throughout Britain excepting only northern Scotland, parts of Ireland and Wales, the south-western

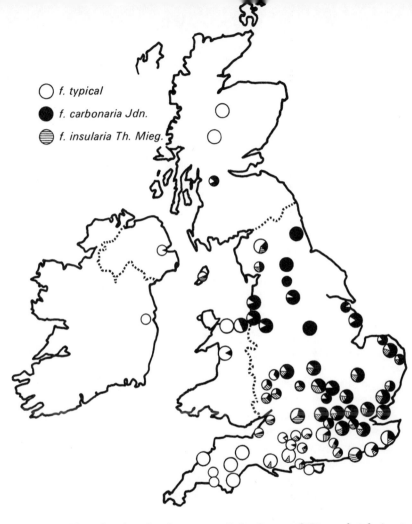

FIG. 4–1. Map showing the frequency of the forms of *Biston betularia* at various sites as determined by H. B. D. Kettlewell and his co-workers. The results are based on the examination of 20,000 specimens. Melanic forms are found in industrial areas and to the east, but not to the west, of them, owing to the drift of pollution by the prevailing westerly winds. (After Ford E. B., 1965, *Ecological Genetics*, 2nd. ed. Map 7).

peninsular and the western end of the south coast. In places it makes up more than 95 per cent of the population. *Insularia* is intermediate in appearance between the typical and *carbonaria* forms. Again it is an industrial melanic, but its distribution differs from *carbonaria* (Fig. 4–1) in that it is more prevalent in the west, for the prevailing south-west wind tends to drift pollution to the east of industrial areas. .

In common with other moths showing industrial melanism *B. betularia* rests on tree trunks during the day, relying on camouflage (see Plate 4) to protect it from predators. Kettlewell has released equal numbers of marked typical and *carbonaria* forms of the moth in both polluted woodlands, near Birmingham, and unpolluted woodlands in Dorset. He was able to show that on the soot-covered trunks of the polluted woodlands there was heavier predation of the typical form by birds, while in the unpolluted woodlands the situation was reversed, the more conspicuous *carbonaria* being more heavily predated.

In our discussion of lichen distribution in Chapter I, the decrease in numbers of lichen species on the various substrates with increase in the level of pollution was noted. It seems likely that the reduction in lichens is a result of the high sulphur dioxide content of the atmosphere. Accompanying the increase in sulphur dioxide is an increase in the amount of soot deposited. The most important factor in the conceal-ment of typcial *B. betularia* individuals is the light-coloured speckled nature of lichen-covered tree bark. Where both the amount of lichen is reduced and the dark deposit of soot increased by surrounding indus-trialization, the typical form becomes conspicuous while dark forms are are less conspicuous. In such situations dark forms are thus at an advantage over light forms and are more likely to survive and reproduce. Natural selection, by bird predation in this case, will thus tend to increase steadily the proportion of dark forms in the population of polluted areas.

Before 1850, as we have said, melanic forms were not caught by entomologists. Presumably they arose by mutation. Previously any such mutants would have been at great disadvantage. In the new situation created by pollution the advantage swung to the side of the melanics. Assuming only that pollution continues at its present level for long enough the typical form will be wiped out except for the occurrence of occasional mutants. Industrial melanism is thus an example of **transient polymorphism**.

We noted above that *insularia* occurs mainly in areas of moderate pollution. Nowhere does it exceed 40 per cent or so of the population. The evidence suggests that there are three allelomorphic genes with *carbonaria* dominant to *insularia*, which is in turn dominant to the typical form. It seems that *insularia* only possesses advantage over *carbonaria* and the typical form in the comparatively short period of time when pollution is building up in a freshly industrialized area or where pollution remains at these levels to the west of industrial areas.

The appearance of a dark form in a population is not necessarily the only effect of a gene for melanism. In some moth species larvae carrying a pair of such genes (homozygotes) or even only one (heterozygotes) are more able to survive severe conditions than homozygotes for the typical form. This advantage could lead to the spread of melanic forms even outside industrial areas unless counteracted by the disadvantage of being conspicuous because they are dark coloured.

Stable polymorphism

Darwin himself studied in detail the heterostyle-homostyle system of the primrose, *Primula vulgaris*, so familiar in elementary biology courses as an example of insect pollination. The majority of primrose plants have either pin-eyed flowers (Fig. 4–2(c)) with a long style and anthers half-way down the corolla tube, or thrum-eyed flowers (Fig. 4–2(b)) with a short style and anthers at the mouth of the corolla tube. Occasionally homostyle flowers occur (Fig. 4–2(a)) with both the style tip, the stigma, and the anthers at the mouth of the corolla tube.

Experiment has shown that pin-eyed is recessive to thrum-eyed. Thrum-eyed pollen, if it falls on a thrum-eyed stigma, may germinate but does not form a pollen tube, and consequently does not fertilize the ovules. Thrum-eyed flowers are therefore all heterozygous. Pin-eyed pollen forms a pollen tube on both pin-eyed and thrum-eyed stigmas. On the former, however, the tube grows more slowly than the tube from

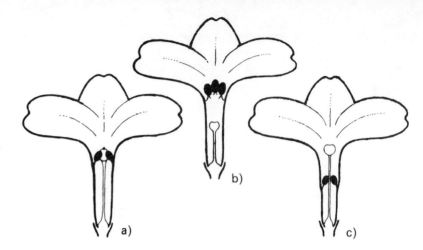

Fig. 4–2. Stable polymorphism. Diagrams of half-flowers of a) homostyle, b) thrum-eyed, and c) pin-eyed Primrose flowers.

thrum-eyed pollen and it is the latter which fertilizes the ovules. Since pin-eyed is recessive all pin-eyed flowers must be homozygous. Because of the genetic control of the two morphological types each plant will produce approximately equal numbers of offspring of both types.

The pin-eyed/thrum-eyed nature of primrose flowers, coupled with the physiological behaviour of the pollen, results in an effective mechanism for ensuring outbreeding. The occasional homostyled primroses which occur as a result of crossing over between the closely linked genes controlling the actual heterostyled flowers have pin-eyed styles and thrum-eyed pollen. They are therefore self-fertile. Because stigma and anthers are so close together homostyle plants are likely to inbreed rather than outbreed. In addition pollen from a homostyle primrose will be as effective as pollen from a thrum-eyed primrose on a pin-eyed stigma. This will result in an increase in the frequency of homostyle flowers. Some of these will be heterozygous, half of them carrying the pin-eyed gene, so that pin-eyed flowers should decrease in frequency less rapidly than thrum-eyed.

Two areas of England, one in Buckinghamshire, the other in Somerset, have populations of almost completely homostyled flowers. Here, instead of the remaining pin-eyed and thrum-eyed flowers being nearly equal in numbers pin-eyed flowers outnumber thrum-eyed by nearly 10 to 1. Why then, if homostyled flowers are at such an apparent advantage have they not spread throughout the country? Why are the majority of primrose populations still maintaining a stable 50 : 50 balance of pin-eyed and thrum-eyed flowers? Ford has suggested that the changing patterns of agriculture and forestry over the past 200 years may have swayed the selective advantage in favour of inbreeding in the two areas in which large numbers of homostyled flowers are found. Only careful experimental growing of homostyled flowers, in various localities and conditions is likely to throw further light on this problem.

Protective colouration

We have already met one example of protective colouration in our discussion of the industrial melanism of *Biston betularia*. Protective colouration by camouflage, or **cryptic colouration**, is exhibited by a large number of animals. For such concealment devices to be effective they must be associated with appropriate behaviour patterns. Peppered moths, for instance, after alighting on a trunk, move slightly to give an approximate alignment of their pattern with that of the lichen covered bark.

Perhaps the most widespread cryptic colouration device is counter-shading. This is to be found in many groups of organisms. It consists simply of a darker colouration of the side which light normally strikes first. In mammals and fish this is usually the dorsal surface, but in the plaice, for example this is altered to fit in with the sideways flattening of the body form during development. Several Hawk moth caterpillars have dark ventral surfaces e.g. the Puss Moth, *Cerura vinula* L. and the Privet Hawk moth, *Spinx ligustri* L. These normally rest on the under-side of twigs and expose the ventral surface to the light. Yet other larvae look like the twigs on which they feed. Even slight differences, as between twigs from different plants of the same species, render conceal-ment less effective and birds are more successful in detecting the larvae.

The intensity of natural selection is particularly great in the case of cryptically coloured insects. A genotype which makes it possible for animals to fit closely to their surroundings is of great advantage. Even movement over small distances on a resting or feeding surface can completely destroy the effectiveness of concealment as further variations in pattern of surface are encountered. In common moth species this results in selection for diversity in colouration.

The stress on careful observation earlier in this book might be applied equally to foraging birds as well as to humans. As with men birds concentrating on a common moth species for food, tend to "get their eye in", and get more expert at finding moths. Variation in cryptic patterns offsets this to some extent and is advantageous to the insects, provided variants from the common pattern are still sufficiently con-cealing. Where cryptic patterns are controlled by single major genes this situation leads to polymorphism as found in *Biston betularia*, where minor additive gene effects are concerned there will be a large number of forms between the possible extremes.

Protective colouration frequently takes the form of making the individual very obvious rather than concealing it, sometimes combining this with initial concealment so bringing in an element of surprise. The eye-spots of many insect species form a second line of defence of this type. At rest they are covered and are suddenly exposed when the insect flies. A. D. Blest carefully investigated this situation and found that not only did insect eye-spots frighten off birds, but that the more like a vertebrate eye they are the more effective they are. This suggests that selection will be for better and better imitations of vertebrate eyes, and it is noticeable that eye-spots originating in differing ways have converged in evolution to a remarkable similarity.

Eye-spots also serve to deflect attention from more vulnerable parts of the body. Butterflies like the Grayling, *Eumenis semele* L. and Small Heath, *Coenonympha pamphilus* L. can still fly with parts of the wing missing. Sheppard records an instance of a Small Heath which was attacked by a lizard at the wing eye-spot.

A sudden alternation of cryptic and bright colour causes confusion to predators. Many insects are well camouflaged at rest but conspicuous on the wing, the sudden disappearance of the bright colour making individuals extremely difficult to spot when they alight.

A further development in conspicuousness is the combination of bright, warning colouration with unpalatable or dangerous features. Once again there are many insect examples, some very familiar, such as the black and yellow banding of wasps in association with their sting, though it also occurs in many other groups. The bright colours of many poisonous lizards and snakes are well known. The preponderance of red, yellow, black and white in warning colouration is noticeable, colours which are often used by humans for the same purpose e.g fire engines.

Mimicry

H. W. Bates, A. R. Wallace, and R. Trimen in the 1860's first showed the occurrence of copying of warning colourations by relatively edible butterfly species throughout the tropics. Since then many further examples have been described. Later F. Müller (1879) pointed out that if two equally objectionable species shared similar colouration, then any predation would fall equally on both of them rather than on one. The essential features of Batesian and Müllerian mimicry may be listed as follows:

Batesian mimicry	Müllerian mimicry
The animal mimicked must be relatively inedible or otherwise protected.	
The animal mimicked must have a conspicuous pattern.	All species are warningly coloured and protected.
The animal mimicked must be common, usually much more common than the mimic.	All species can be equally common.
Both the model and the mimic must usually be found in the same area at the same time.	
The mimic must closely resemble the model.	Resemblance between the species need not be very exact.
The resemblance is only visual: colour, pattern or behaviour.	
	Species are rarely polymorphic.

As the above list would suggest it is not always easy to decide whether a particular mimic is Batesian or Müllerian. Indeed a mimic could be a Batesian mimic of one species and a Müllerian mimic of another. Mimicry of these two types is, once more, particularly well shown by arthropods, especially spiders and insects.

Mimicry lends itself to experimental test. Brower used starlings, *Sturnus vulgaris* L., as predators and meal-worms as prey. Models were made distasteful by dipping in 66 per cent quinine dihydrochloride and were distinguished visually by a band of green paint. Other meal worms were either dipped in distilled water and banded green to make mimics, or banded orange to determine whether the birds could learn to associate what is commonly a warning colour with edibility. Control birds were fed mimics only, which they ate, showing the green paint did not of itself result in aversion.

Nine birds were used for a total of 160 trials consisting of one green banded meal worm and one orange banded meal worm according to the following scheme:

Group A (four birds)		
Green banded		Orange banded
% Model	% Mimic	
(control) —	100	+
90	10	+
70	30	+
—	—	—
10	90	+

Group B (five birds)

| Green banded | | Orange banded |
% Mode	% Mimic	
—	100	+
90	10	+
70	30	+
40	60	+
10	90	+

In all trials eight of the nine starlings ate all the orange banded edible meal worms, while the ninth receiving 70 per cent models and 30 per cent mimics refused meal-worms after the sixty second trial.

After first testing the quinine dipped models, the birds avoided them, usually recognizing them on sight. In consequence the mimics also proved to be protected by the green banding. While mimics made up 10, 30 and 60 per cent of all green banded meal worms about 80 per cent of them remained uneaten. Even the two birds receiving 90 per cent mimics left 17 per cent of them alone. Birds varied in their ability to learn and to remember the association of colour with edibility. Most of the starlings occasionally made mistakes and took models after previously learning to avoid them. Brower points out that this allows the predator to detect changes in the relative numbers of models and mimics and is of importance to it.

Experimental work, such as that described above, makes it possible to reach a reasoned understanding of the evolution of both Batesian and Müllerian mimicry. It is an interesting exercise in close argument to use the information gained by Brower and the features of mimicry listed on page 22 to follow through the possible evolutionary sequence resulting in Müllerian mimicry of two equally distasteful species, A and B, inhabiting the same area and having a small degree of initial resemblance to each other. This, and other posssibilities are discussed by Sheppard.

Before leaving the subject of mimicry, the mimicry of insects by flowers is worth mentioning as an example that can, of course, be neither Batesian nor Müllerian. In Britain the bee orchis, *Ophrys apifera.* Huds., and fly orchis *O. insectifera* L. have flower parts which resemble the female abdomen of the bumble-bee in the former case, and are more generally bee like in the latter. Male insects are attracted by both the scent and appearance of the flowers and dislodge pollen when attempting to copulate with them. The pollen is carried to another flower of the same species since each orchid species attracts only the males of the insect species the females of which are mimicked by its flowers.

Plant ecotypes

The Swedish botanist Turesson noted that the majority of plant species ranged through several habitats, and often through several climatic zones also, and that populations in different parts of the range showed adaptations of physiology and form. Samples of these variations were collected and grown under uniform conditions in a neutral area. In a few cases he found that very varied forms grew to be similar after only a few seasons under cultivation in the neutral situation. Usually, however, differences between plants from different habitats were persistent and he was able to divide species into ecotypes adapted to their own particular habitats.

Amongst many later studies that of the sea plantain, *Plantago maritima*, by Gregor and his co-workers in Scotland has especially emphasized that ecotypes may frequently be less distinct that Turesson supposed. The distribution of Plantago maritima shows a continuous gradation in variation, a cline, and is made up of more or less isolated demes, or breeding poulations (Fig. 4–3).

Adaptation

The instances of evolution in action given in this chapter are only a small proportion of those so far studied. Adaptation to environment results from the operation of natural selection. The familiar text-book illustration of this referring to Darwin's investigation of the finches of the Galapagos Islands has, however, been supplemented by many examples of the investigation of adaptation occurring now, and nearer home. Nevertheless, we have barely begun the study of the operation of the two major forces of evolution, natural selection and mutation.

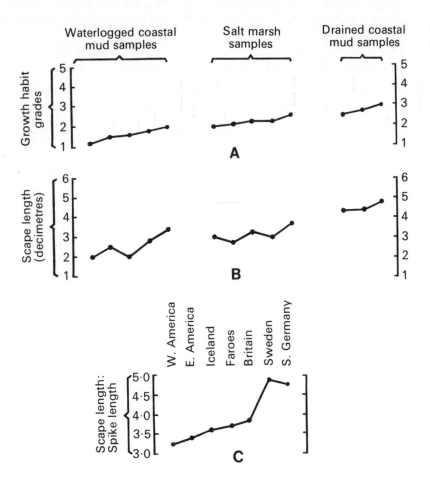

FIG. 4–3. Clines shown by *Plantago maritima*. A. An ecocline in growth habit. B. An ecocline in scape (flower stalk) length. C. A cline within the North Atlantic region of the ratio scape length to spike length. (AfterHeslop-Harrison J. 1953 *New Concepts in Flowering Plant Taxonomy* Fig. 5, p. 55, Heinemann).

Summary

1. The combined field and laboratory study of evolution in action has been termed Ecological Genetics by Ford. This chapter describes some aspects only.

2. Industrial melanism in the peppered moth is described as an example of transient polymorphism, while the heterostyle-homostyle system in the primrose exemplifies stable polymorphism.

3. Protective colouration includes both cryptic patterns and bright patterns which deflect attention.

4. Mimicry is discussed at some length. This, and work on plant ecotypes, illustrates the use of experiment in ecology.

5. Adaptation of organisms results from the operation of natural selection, the study of which, partly by methods discussed in this chapter, has only just begun.

Chapter V
Concepts in ecology

THE USE of the basic approaches to ecology described in Part I has uncovered relationships within the living environment and between it and the physical environment. Some relationships are constant enough in their outlines to have made it possible to derive ecological concepts from them, which can be used to predict either the existence of similar relationships in other environments or a likely future sequence of changes in relationships. Because ecology is a scientific discipline the concepts used as an aid to investigation are dynamic not immutable. They must be capable of change in response to the acquisition of new knowledge, and continuous modification in the light of their usefulness.

You will see that we have already discussed many of these concepts in previous chapters. Others have grown from an application of the concept that the behaviour of a **population** of individual organisms is something more than a simple sum of the behaviour of its individuals. Because, in the field at least, total populations are rarely accessible to an investigator, work on them relies heavily on extensions of the numerical approach to ecology. Slowly ecologists are developing mathematical statements of their working concepts. This process is only in its infancy and largely confined to research.

Biology can be subdivided in many ways. Fig. 5–1 illustrates one such division based on the degree of organization of the units.

BIOLOGY

(1) Sub-cellular   Biosphere (10)
  (2) Cell   Ecosystems (9)
    (3) Tissues   Communities (8)
      (4) Organs   Populations (7)
        (5) Systems of Organs   Individual Organism (6)

Fig. 5–1 The organizational divisions of biology.

In this diagram the right-hand half, units 6, 7, 8, 9 and 10 cover the range of ecological study. It is essential to remember that what goes on inside individual organisms is important and affects their behaviour within populations for all units interact with each other. The more interaction there is the nearer together they have been placed in the diagram. Possibly the most useful ecological concept in the initial stages of study is the **ecosystem**—the complex of living and inert materials which make up a more or less balanced unit. Most ecosystems are moderately stable in terms of the life span of man and so have common names—lake, river, wood, sea-shore, are all ecosystems. Within an ecosystem it is useful to consider the feeding of organisms, and therefore how interchange of energy within the ecosystem, goes on. Ecosystems may be subdivided into

1. **Abiotic** (non-living) **environment**, *interacting with*
2. **Biotic** (living) **environment**
  (a) **Autotrophs** (The **producers**, requiring only the input of light energy and simple inorganic salts), *interacting with*
  (b) **Heterotrophs**
    a) The **consumers**, (Deriving energy direct from autotroph substance—**herbivores**—or indirectly through earlier consumers—**carnivores**).
    b) **Decomposers** (Deriving energy from the decomposition of the remains of both autotrophs and other heterotrophs).

The interaction of autotrophs with heterotrophs is a feature of all ecosystems. Interaction may not be immediate. In woodland for example there is a partial separation in space, heterotrophs being concentrated on the woodland floor, and a separation in time, since decomposers must await leaf, twig or branch fall before the energy stored can be utilized. Within an ecosystem, such as a woodland, it is valuable to be able to separate off smaller geographical units, or habitats, where different processes go on or groups of closely interrelated organisms live, e.g. the tree canopy, rotting logs. The larger habitats within an ecosystem are sometimes profitably thought of as separate, interacting **communities** each containing organisms capable of most of the variations of energy use possible but not quite so self-sufficient as the whole ecosystem. Communities can be thought of as structured in space, as habitats or according to variety of method in energy use, as made up of ecological **niches**. Each species plays a particular role within the community and it is noticeable that no two quite overlap. Even if the job is the same, as for example, predation of small mammals by hawks and owls, it is separated in time.

The interactions between organisms, and between the biotic and abiotic environments, provide a system of checks and balances which results in the continued existence of ecosystems. When these **homeostatic mechanisms** break down perhaps as a result of pollution by man, the result may be complete destruction of an ecosystem or such great changes that balance takes a long time to restore.

The whole series of energy changes going on in the **biosphere**, the region above, on, and a little below the earth's surface in which life as we know it can exist, result from the input of light energy which, has no foreseeable limit. The chemical elements contained within the biosphere do, however, obviously have a finite limit and if all or any one of them, required for life, were to be entirely converted first to autotroph substance, then to heterotroph substance, and to stop there, life would end. At this stage, of course, the decomposers return the elements to their inorganic state and they become once more available for autotrophs. To this cycling of the elements through living organisms must be added cycling under abiotic conditions as a result of rock weathering and climate. The resulting **biogeochemical cycles**, e.g. the nitrogen cycle, are a delicate part of the ecosystem. Man's operations frequently tend to lock up materials in part of the cycle and transfer elsewhere is delayed too long for the ecosystem to remain stable.

At the beginning of Chapter 3 we noted that **food webs** are built of **trophic levels** always commencing with a producer which is in turn followed by one or more levels of **consumers**. A food web is, in fact, another way of expressing the feeding relations within an ecosystem, listed on p. 25, in terms of specific organisms. We saw also that, within communities, it was possible to represent the trophic levels by **pyramids of number**, **biomass**, or **energy**, the latter being the most satisfactory, although the hardest to determine.

A constant feature of the energetic study of ecosystems is that the rate at which energy passes through each trophic level, the **energy flow** is always less than through the preceding level. This follows from the second law of thermodynamics since the efficiency of energy intake to each level is depleted by heat loss through respiration, and the resulting **ecological efficiency** must always be less than 100 per cent. At any one moment in time each trophic level will contain a certain amount of energy, the **standing crop**. This is distinct from the rate of storage of energy in living tissues during a given period of time, the **nett productivity**. As in the example given in Fig. 3–1(e) for biomass it is possible that the standing crop energy could be a great deal lower than the nett productivity. This is likely to be so in the case of small organisms with high metabolic rates. Energy lost in heat of respiration is consequently high with a resulting low standing crop energy although nett productivity may be as high as, or higher, than for larger organisms.

As early as 1840, Justus von Liebig, working on the inorganic materials needed for plant growth had noted that the substance most nearly approaching the tolerable minimum set a limit to growth. By 1913 much work had been done in this field and the idea of **limiting factors** was extended by V. E. Shelford to cover many possibilities besides plant mineral nutrients and refined to a concept of **limits of tolerance**. The growth of organisms is limited not only by deficiency in the amount or quality of a range of interacting factors, but also by excess. The range of tolerance of organisms to temperature, radiation, water, nutrients, soil conditions, etc. is not necessarily of the same degree for all factors. Uniformly wide tolerance ranges are likely to result in the wide distribution of an organism.

A **population** of organisms is a group of organisms of the same species occupying a particular space. Such a collective grouping has properties which the single individual cannot have—density, a rate of birth, a rate of death, a distribution of ages amongst the individuals of the population, and so on. The study of the properties of populations depends upon the use of sampling, statistical methods, and mathematical models.

Populations of organisms occupy a certain amount of space. The numbers of organisms or their biomass in a unit of space give a value for the **density** of the population. Frequently the area concerned is so large or the irregularity of the **distribution** of organisms is such that a measurement of **relative abundance**, e.g. the percentage of samples in which an organism occurs, is a more convenient measurement of population relationships with area, volume or time. The **micro-environment** or **microhabitat**, the immediate surroundings in which a species can live, is one factor producing clumping and irregular **dispersion** within communities.

The **growth form** of a population is compounded of its **birth rate** (or rate of immigration) and its **death rate** (or rate of emigration). Where no checks exist to the growth of a population a graph of numbers of individuals against time—the **growth rate**—takes the form shown in Fig. 5–2(a). Imagine, for example, a single bacterium placed in a culture medium of unlimited volume and constant nutrient balance, which is capable of division once each half-hour. From a single individual at midnight by 8 a.m. there would be 65,136 individuals since growth would be by simple geometric progression 2, 4, 8, 16, 32, and so on. It is clear that this is not a common situation or the biosphere would have become a solid mass of bacteria long ago. Checks do, in fact, operate. Space is limited, or food is limited, or other factors set a limit on growth. For the majority of populations the form of growth is as in Fig. 5–2(b). Initial geometric increase is limited and the population comes into balance with its environment.

P-F. Verhulst has constructed a mathematical model of population growth. He suggested that (a) When growth rate is unchecked the rate is directly, and constantly related to the size of the population (Fig.

5–2(a) and (b)) When growth rate is checked it is reduced by a factor which varies directly with the square of the size of the population (Fig. 5–2(b) and (c)). Thus

when R_n = rate of growth of an unchecked population
N = number of individuals in the population

then $R_n = k_n N$ or $\dfrac{R_n}{N} = k_n$ $\qquad$ (1)

where k_n = the constant relating R_n and N.

k_n measures the ability of the population to increase and is known as its **biotic potential**.

Also

when R_c = rate at which the growth of a population is checked
N^2 = square of the number of individuals in the population

then $R_c = k_c N^2$ or $\dfrac{R_c}{N^2} = k_c$ $\qquad$ (2)

where k_c = the constant relating R_c and N^2. k_c measures the resistance to its own growth created by a growing population and is known as its **environmental resistance**.

Equations (1) and (2) express the two parts of Verhulst's hypothesis. Since the rate at which a population is able to grow is a combination of these, (1) being a factor for increase and (2) a factor for decrease, we can write

$$R = R_n - R_c$$

where R = actual rate of growth of a population. By substitution we can then write

$$R = k_n N - K^c N^2 \qquad (3)$$

Equation (3) represents a final mathematical model for population growth which can be used for further studies.

Populations showing the S-shaped growth form illustrated in Fig 5–2(b) are brought gradually to an upper limit or **carrying capacity**

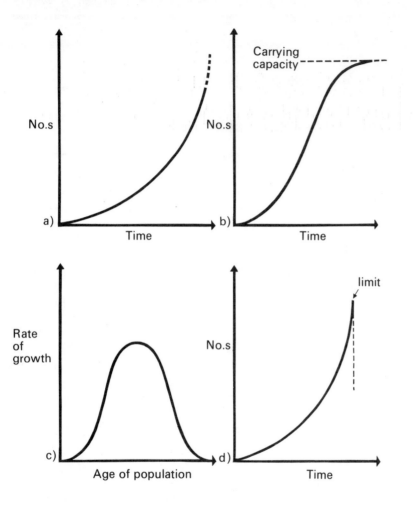

FIG. 5–2. The growth of populations: a) Unchecked growth, b) Limited growth, c) Rate of growth curve of a population of limited growth, d) Sudden imposition of environmental resistance.

Where environmental resistance suddenly increases with increase in density a J-shaped curve, as in Fig. 5–2(d) results. Here no equilibrium develops. Even where populations tend to maintain a level of carrying capacity there is often fluctuation about this level which may be regular and cyclic. The small amount of work done indicates that energy flow measurements provide a more reliable interpretation of population fluctuations than number and biomass.

The **age distribution** of a population influences both the birth rate and death rate, the latter usually varying with age, and reproduction often being confined to restricted age groups. The relative proportions of pre-reproductive, reproductive and post-reproductive individuals in a population are important to its maintenance. Expanding populations contain large numbers of young individuals, while declining populations contain large numbers of old individuals. As with other population characteristics, age distribution proportions tend to fluctuate around a level which maintains stability.

Dispersal of organisms may mean **emigration** from an area, **immigration** to it or migration within it. As we have already noted dispersal is an additional factor affecting population growth form. Dispersal is much affected by barriers (see Chapter I p. 6) and the **vagility**, or potential for movement, of individuals.

All factors, whether tending to increase the growth rate of a population or to decrease it must either act regardless of the density of the population i.e. **density independent factors** as, for example, severe climatic changes, or vary with the population density i.e. **density dependent factors**. It is particularly noticeable that density independent factors affect smaller organisms most, while larger organisms are particularly affected by such density dependent factors as intraspecific competition and interspecific predation.

The aggregation of individuals within populations in response to microenvironment, weather changes, reproductive needs, or social attraction is commonplace. Perhaps less common are forces resulting in the spacing out of individuals. This usually results from inter-individual competition or mutual antagonism. In many vertebrates and arthropods individuals, pairs or family groups develop a more or less

TABLE VIII

Two-species population interactions

| Type of interaction | Effect on population growth and survival of two populations, A and B | | | | General result of interaction |
| | When not interacting | | When interacting | | |
	A	B	A	B	
1. Neutralism (A and B independent)	0	0	0	0	Neither population affects the other
2. Competition (A and B competitors)	0	0	—	—	Population most affected eliminated from niche
3. Mutualism (A and B partners or symbionts)	—	—	+	+	Interation obligatory for both
4. Protocooperation (A and B cooperators)	0	0	+	+	Interaction favourable to both, but not obligatory
5. Commensalism (A commensal; B host)	—	0	+	0	Obligatory for A; B not affected
6. Amensalism (A amensal; B inhibitor or antibiotic)	0	0	—	0	A inhibited B not affected
7. Parasitism (A parasite; B host) 8. Predation (A predator; B prey)	—	0	+	—	Obligatory for A; B inhibited

+ Population growth increased   — Population growth decreased   0 Population growth not affected

After E. P. Odum 1959 courtesy W. B. Saunders Co.

defined **territory**, which reduces competition and overcrowding.

Interactions between populations of more than one species vary, from no interaction at all, at one end of the scale, to complete dependence of one population on another. These interactions are summarized in Table VIII. Most are familiar from elementary biology courses. It is important to remember that the impression of relationships between individuals is a false one and that symbiosis, parasitism, and so on are actually population interactions.

A group of populations living in a particular area or habitat makes up a **biotic community**. A community tends to have a relatively uniform appearance, constancy of species composition, trophic organization and pattern of energy flow. Certain organisms within the community tend to control the energy flow and are ecological **dominants**. Since these, in terrestrial communities at least, are often visually striking also they have often been used as a basis of community classification. Such a classification has the disadvantage that either the visually obvious organisms may not, in fact, be the ecological dominants or several organisms may be equally prominent. Populations overlap in space and time with the result that community boundaries are not always easy to draw and they are best thought of as being distributed along a gradient or continuum.

In any area a stable **climax** community cannot suddenly come into existence. The climax develops as a **sere**. The sequence of communities, or **seral stages**, moves towards a uniform community throughout the area. Newly exposed areas, such as a rock face, develop **primary**

successions of organisms. When these successions are removed, by change in the factors tending towards the climax at whatever seral stage, a **secondary succession** can develop.

At the junctions of very different communities, as between a wood and a pond or grassland, there is a transition zone which it is sometimes helpful to separate from either community as an **ecotone**. In such areas there is a likelihood of increased variety of organisms and greater densities.

Change within a community is not only of the continuing successional type but may be periodic. **Periodicity** may be daily, seasonal, lunar (as with the tides), or inherent in the organisms themselves. Particular attention has been paid to the latter in the study of the physiological synchromization of activities with periodic environmental changes by so-called "biological clocks".

Using the same argument as the geologist observing weathering processes today, interpreting past geological activities in terms of the same processes, the ecologist may interpret evidence of past ecological events. **Palaeoecology** has proved an especially fruitful technique in post-glacial deposits, indicating considerable past climatic changes.

Summary

1. Relationships between ecological concepts are described.
2. The majority of ecological concepts apply to populations, communities, and ecosystems, and not to individual organisms.

PLATE 1. Urban and rural environment. The whole of this landscape has been affected by man. On the one hand the tall blocks of flats, terraced houses and pavements provide suitable habitats for a variety of rodents, insects, grasses and many other organisms. On the other hand the fields and moorlands in the background are coated with a film of soot from the town and many plant species, and consequently animals, have disappeared from this area in the last one hundred and fifty years.

(*Photograph: J. Nicholson*)

31

PLATE 2. Microhabitat. *A*. A fresh cow-pat with flies at rest on the surface. *B*. An older cow-pat. The surface is now dried and cracked. No flies visit it but beneath the surface a succession of insect larvae, worms and fungi live) (*Photograph: J. Nicholson.*

PLATE 3. Succession. Part of the sand-dune succession Ainsdale National Nature Reserve, Lancashire. In the centre is the beginning of a large blow-out caused by wind following the exposure of the sand surface by the trampling feet of visitors. The pine woods in the background have been planted on the older more stable dunes.

(*Photograph: J. Nicholson*)

PLATE 4. Polymorphism. *A*. Peppered moths, one typical and one *carbonaria* form resting on lichen covered bark in an unpolluted area. (Approximately half natural size).

B. Peppered moths, one typical and one *carbonaria* form resting on the bark of a tree in an industrial area. (Approximately half natural size.)

(*Photographs: J. Haywood*, after Ford E.B.*Ecological Genetics*, plates 14 and 15)

Chapter VI

PART II

Plant ecology

THE BASIC objects of any description of vegetation are to record the present state of the vegetation, and to provide a record which others can understand and which can be compared with records from other areas. Recording the present state of the vegetation is largely a matter of field observation and is divisible into three main aspects.

1. **Floristic composition**—a list of the species of plants present.
2. **Life-forms**—What sorts of plants are present.

The Danish botanist, Raunkiaer, devised a classification of plants based on the height of the buds which enabled a plant to live through adverse seasons. A summary of this classification is given in Table IX.

TABLE IX

The life forms of plants.

		Position of over-wintering buds or shoot apices.	Examples
Phanerophytes	a) Evergreen b) Deciduous	on aerial shoots close to the ground.	Scots Pine Oak
Chamaephytes Hemicryptophytes Cryptophytes		At ground level Below ground or submerged in water.	Crowberry Daisy
	a) Geophytes b) Helophytes	(Bulbs or tubers) In soil or mud below water.	Daffodil Waterplantain
	c) Hydrophytes	In water, floating or submerged leaves.	Water lily
Therophytes		None. Annuals.	Shepherds' Purse

3. **Structure**.

a) *STRATIFICATION*.

When one enters a woodland, it is apparent that the plants within it form at least two vertically distributed layers, the tree canopy and the ground vegetation. In many cases there are further layers distinguishable. Lower vegetation e.g. grassland and heathland can be looked at in this way. (Fig. 6–1). There is a danger that different investigators may decide to draw the boundaries between layers at different heights and it is therefore important to state the height boundaries used.

b) *DISTRIBUTION*.

The mapping of individual plants in detail is both laborious and time-consuming. Usually the return for the amount of effort put in is low. Permanent plots, frequently mapped over months or years, are however sometimes useful, but there is often considerable interference with them in areas to which the public has access. Comparison of the resulting maps must be by eye and is consequently ineffective in the detection of small slow changes.

c) *ABUNDANCE*.

1. **Subjective measurement of abundance**.

A. *Frequency symbols*.

Each species within the list for an area can be quickly visually assessed for its abundance relative to others present, and divided into classes e.g. Dominant, Abundant, Frequent, Occasional, and Rare. These can be further subdivided if required.

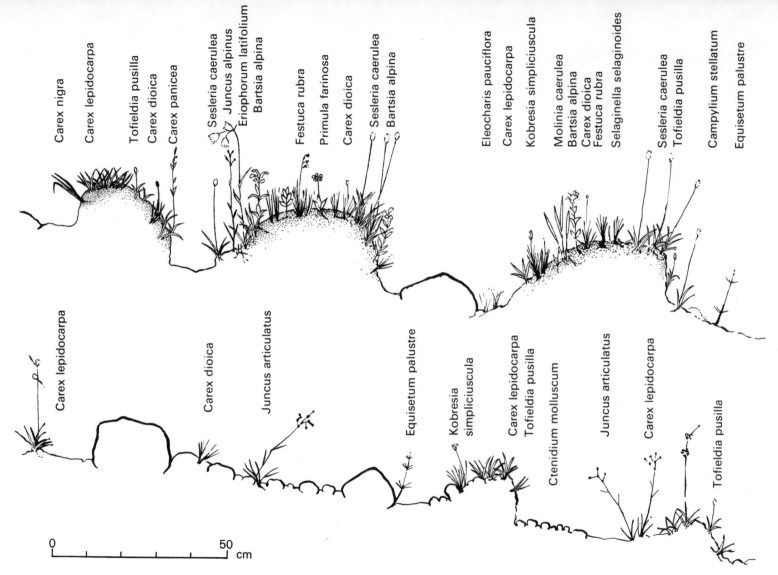

Carex nigra
Carex lepidocarpa
Tofieldia pusilla
Carex dioica
Carex panicea
Sesleria caerulea
Juncus alpinus
Eriophorum latifolium
Bartsia alpina
Festuca rubra
Primula farinosa
Carex dioica
Sesleria caerulea
Bartsia alpina
Eleocharis pauciflora
Carex lepidocarpa
Kobresia simpliciuscula
Molinia caerulea
Bartsia alpina
Carex dioica
Festuca rubra
Selaginella selaginoides
Sesleria caerulea
Tofieldia pusilla
Campylium stellatum
Equisetum palustre

Carex lepidocarpa
Carex dioica
Juncus articulatus
Equisetum palustre
Kobresia simpliciuscula
Carex lepidocarpa
Tofieldia pusilla
Ctenidium molluscum
Juncus articulatus
Carex lepidocarpa
Tofieldia pusilla

0   50
| | | | | | | cm

FIG. 6–1. A profile diagram through the hummock complex at the upper margin of a small calcareous marsh. The lower diagram is a direct continuation of the upper. (After Pigott C. D., 1956, *J. Ecol.* p. 545–586).

36

Subjective recording of this sort is strongly affected by the conspicuousness of a plant species, in both colour and size, by its spatial distribution, and by the experience of the recorder. Even for those with long experience of using frequency classes for recording, the margin of error remains large.

B. *Braun-Blanquet system.*

A considerable refinement of the subjective assessment of abundance has resulted from the work of Professor J. Braun-Blanquet. The Domin modification of the original scale (Table X) can be applied to a series of adjacent quadrats of increasing size, laid down as shown in Fig. 6–2.

This method is rapid and particularly suited to the initial investigation of an area. Because the size of plants and their distribution

are not taken into account it is easier for individual observers to reduce errors. It still suffers from the defect that it is difficult to get agreement between different observers.

2. **Quantitative measurement of abundance**.

The object of quantitative measurement is to reduce error to an amount which can be estimated statistically. Inevitably this means a longer time spent in field measurement and which measure of abundance should be used in a particular case is often decided by this.

A. *Density.*

This is a count of the number of individual plants within a quadrat. The quadrats should be randomly placed (see Chapter II p. 10). Large numbers of plants may be involved, and it is sometimes difficult to define what is meant by an "individual" as, for example, with the grasses.

B. *Cover.*

The percentage of the total area covered by each species is estimated. Since at any one point leaves of more than one plant may overlap the total of individual species percentages may well be more than 100 per cent.

If quadrats are used the percentage cover has to be estimated visually. This estimation is likely to include the same subjective

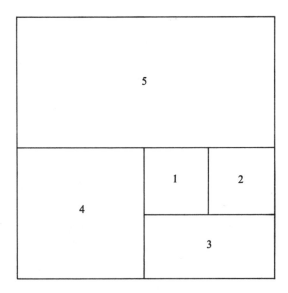

FIG. 6–2. Successive increase in size of the area sampled by an arrangement of adjacent quadrats.

errors already mentioned. Error can be cut down by sub-dividing the quadrat into smaller squares and estimating within each. The ultimate logic of this is to use points, for plant parts are either touched by the points or not, either 100 per cent in the sample or 100 per cent out of it. A point is most easily obtained by arranging a frame above the vegetation on which optical cross-wires are supported. Simpler to produce is a frame through which a metal pin can be dropped. There is a statistically significant difference between values for cover obtained by using a metal pin of definite diameter, and those obtained as absolute values using optical cross-wires. Provided the investigation is comparative between one site and another this does not matter. Several pins may be placed in a single frame and the position of the whole frame randomized for sampling. Individual pins will not be randomly placed using this method. For comparative purposes this is again unimportant.

C. *Frequency.*

Instead of visually assessing the abundance of a species in an area the percentage of randomly placed quadrats in which it occurs is recorded. Recording must be consistently either only of plants actually rooted within the quadrat, or of plants with aerial parts within the quadrat.

The frequency value obtained is greatly affected by

(i) The size of the quadrat. The larger the quadrat is the higher the frequency will be and the quadrat size should, therefore, always be stated.

(ii) The size of the plant. When aerial plant parts are recorded large plants register higher values. Only the recording of rooted plants removes this bias.

(iii) The spatial distribution of individuals. Plants grouped in clumps are likely to be missed by relatively small, randomly placed quadrats.

Although rapidly determined frequency is too subject to error to be useful except in large-scale work.

D. *Yield.*

Clippings from sample quadrats are sorted into species, dried, and weighed. This is a good measure of abundance but is very time-consuming.

E. *Performance.*

Measurements of leaf width, leaf length, the ratio of the two, flower number, seeds per capsule etc. all indicate how efficiently a plant is performing in its environment. Such measurements are particularly useful in autecological studies in small areas.

Understanding and comparing records

Understanding records is important for the recorder as well as for those who wish to refer to his work later. We have already discussed the need for accurate written, and other, records in Chapter I. It is necessary, once more, to re-emphasize this for as soon as numerical data are taken accuracy and completeness require greater care to achieve. Always remember that complete records include a full account of how they were obtained. Without this they may well be useless for further analysis or purposes of comparison.

Comparison of records, both within a single investigation, and between investigations requires careful planning if it is to be valid. In Chapter II we noted the need for sampling large populations and some of the ways in which this may be done. The correct way to sample is usually derivable, with careful thought, from the nature of the investigation, the information required, and the statistical tests to be used. Work in the field often takes up a lot of time so it is doubly important to both sample correctly, and sample sufficiently, so that a second visit for further information is avoided: over rather than under sample.

Some statistical methods have been described in Chapter II, and the Appendix, contains a resumé of these and others. How is information to be collected in the field, and which statistical tests may be used on the data, and for what purposes?

Field methods for the objective sampling of vegetation may be divided into:—

1. *Quadrats.* Usually these are square frames, of a size suited to the investigation. Any materials may be used but adjustable, folding wooden or metal ones save effort, especially over difficult terrain.

2. *Point frames.* These are in effect the minimal quadrat. The smaller the diameter of the pin the better, and the fewer pins per frame the better. Knitting needles in frames of ten are a reasonable compromise in most circumstances.

3. *Transects.* Quadrats, or point frames, are placed at known intervals along a line across an environmental gradient to be investigated. A variation of technique is to divide an environmental gradient into zones and to enumerate a given number of quadrats within each zone.

4. *Isonomes.* The study area is enumerated by a grid of adjacent quadrats. The measure of abundance used is then plotted into squared paper and those squares with approximately equal values joined by lines to give a "contour" map of abundance. The same process repeated for an environmental factor such as soil moisture gives a similar map which can be compared with the first as an overlay.

These Field methods and some methods of analysis are summarized in Table XI. It is beyond the scope of this book to consider the analysis of the detailed pattern of plant communities. Discussion of these is to be found in the more advanced texts listed in the Bibliography.

The classification of plant communities.

In Chapter V p. 29 we discussed the ecological concepts of plant communities which have been built up as a result of the application of the above, and more advanced methods, of plant ecology. It is in the nature of the human mind to understand the surrounding world by simplifying and classifying into sub-units. Investigation of plant communities using the methods of plant ecology will help you to understand these concepts more fully and lead you to think out the reasons behind them, to classify for yourself and, more importantly, to use plant ecological concepts correctly.

TABLE XI

Summary of quantitative field methods in plant ecology and their analysis.

Field Method	Types of sampling	Method of comparison	Use
Quadrat or Point-frame	Random	't' test	Deciding if a difference between two sets of data is significant.
		χ^2 test	Association between spp. Simple presence or absence data with some quadrats containing neither spp. (N.B. this depends on quadrat size).
		Scatter diagram & correlation coefficient	Association between two spp. or abundance measure of a sp. and measure of environmental factor (N.B. depends on quadrat size).
		Regression analysis	Calcn. of equation best fitting correlation data.
Transect	Regular	—	Change of vegtn. along an environmental gradient.
Isonome	Regular	—	Distribution of sp. in relation to variation in environmental factor.

Summary

1. The description of vegetation may be considered under three headings

    a) floristic composition

    b) life-forms

    c) structure.

2. Methods of describing the structure of vegetation are discussed, with special consideration of quantitative measures of abundance.

Chapter VII

Animal ecology

THE development of techniques for the study of the ecology of animals has been governed particularly by two factors. Firstly animals move. Most animals are capable of movement throughout their lives, while others are sedentary for part of the time, usually in the adult phases. Secondly, and greatly influencing the difficulties resulting from movement, the variety of size in animals is very great ranging from microscopic to elephantine.

As a result of these two factors, the development of animal ecology techniques has been uneven and there has been a tendency for workers to concentrate on groups or sub-groups of animals e.g. insects, or small mammals. In consequence the survey contained in this chapter cannot have the cohesion of the survey of plant ecology techniques contained in Chapter VI.

Animal observation

Not only the movements of animals, but their size relative to humans, plays an important part in the technique of animal observation. First efforts are often frustrated by a lack of knowledge of the behaviour patterns of the species under study, and only trial and error can remove the frustration. Insects are particularly susceptible to clumsy movement and shadowing, while light reflection from hands and face often scares off a bird. Smooth, slow, quiet movement only comes with practice, but is a habit worth cultivating even though, in some instances it may be of no direct value.

Records of animal observations have to be surprisingly detailed to be of later value. Notes such as, "the female fed the chick," are almost useless. Did the chick elicit the feeding in any way? Did the adult regurgitate the food? Is the pattern of movements before, during and after feeding a constantly repeated one, or does it vary? Answering these and similar questions requires care and patience.

Observation of animals under their normal habitat conditions is an obvious requirement for the ecologist. Where this entails activity at night, or in complete darkness as in caves, the human observer must inevitably modify the environment with additional light if he is to observe directly. Under such conditions and in situations where the pattern of animal activity is not convenient for humans to follow, it is helpful to turn to the second-hand evidence provided by electronic and mechanical instruments. Similar reasons make instruments useful for recording change in physical habitat factors such as temperature and humidity.

Absolute estimates of animal numbers

It is possible to take samples for the calculation of absolute estimates of animal numbers in several ways. The two most generally useful methods are described first.

1. **Capture—Recapture Techniques**.

MARKING THE ANIMALS.

It is important that techniques used for marking animals should affect neither their health nor their behaviour. Considerable work has already been done on these aspects of marking and reference should be made to books mentioned in the bibliography for further details. Amongst materials used for marking are:—

(A) *Paints and dyes in solution*. These are especially useful for insects and have been used on other groups also. There is a particular need to take care that the solvents used are non-toxic.

(*B*) *Dyes as powders.* Some specialized applications of these have been successful but they are not generally useful.

(*C*) *Labels including rings.* Bird ringing with both numbered and coloured rings is extremely efficient, while mammals of various species have been successfully tagged. It is a basic requirement for all label types that the organism should be relatively large, and the method has consequently been of little use for insects.

(*D*) *Mutilation.* Removal of parts of ear, fin, toes, elytra, or wings can be useful where observation is sufficiently close for this to be easily visible. There is an inherent danger of harm to the animal and the method is best avoided whenever possible.

(*E*) *Internal marking with dyes.* Attempts to do this by injection or feeding remain specialist so far.

(*F*) *Radioactive tracers.* Radioactive cobalt (Co[60]), Tantalum (Ta[182]), phosphorus (P[32]), and carbon (C[14]) are amongst many isotopes which have been used. The method has been extensively used for insects, and in other groups as well. Detection is by a Geiger-Müller tube or by autoradiography. The use of radioactive isotopes in educational establishments in Britain is under strict control by the Department of Education and Science and a detailed check on requirements should be made before plans are laid to use them.

RELEASE

When it is necessary to cage animals for working, subsequent release should be carefully controlled. Animals with a definite periodicity in activity are best released during their least active period. Sites for release should be scattered throughout the habitat to ensure thorough mixing of the marked animals with the unmarked population.

ANALYSIS OF CAPTURE—RECAPTURE DATA

The basic principle of the capture—recapture technique is that if a proportion of a population is marked, returned to the original population and then, after complete mixing, a second sample is taken, the number of marked individuals in the second sample would have the same ratio to the total numbers in the second sample as the total of marked individuals originally released would have to the total population. As the first three values are known the last can be calculated. Four assumptions underlie the application of the method.

(A) That the marked animals are unaffected by the marks and that the marks will remain on all individuals throughout the investigation.

(B) That the marked animals become completely mixed throughout the population.

(C) That the population is sampled randomly so that

1. The different age groups and both sexes are sampled in the proportions in which they exist in the total population.
2. All individuals are equally available for capture irrespective of their location in the habitat.

(D) That sampling is at discrete intervals of time and that sampling time is small in relation to the total time. The problems arising under (A) and (B) have already been discussed. Further discussion of (C) and (D) will be found in Southwood.

The basic formula for the interpretation of capture-recapture data is often known as the "Lincoln Index" after its American originator. It is applicable to a population of fixed size or a population which is subject to either gain or loss but not to both. From the basic principle given above we can write

$$\frac{R}{N} = \frac{m}{S}$$

Where R = number of marked animals in the population
       N = population size
       m = number of marked animals in recapture sample
       S = number of animals in random recapture sample.

It will be obvious that the "Lincoln Index" very much over simplifies the situation in the field. To try to overcome this disadvantage the formula has been further developed by Bailey, Fisher and Ford and Jolly. Parr, Gaskell and George (see bibliography) have summarized these methods.

It is important to choose a formula appropriate to the particular investigation and to specify exactly all the known limitations in the interpretation of results.

2. Unit Sampling of Habitats.

The principal of this method is to sample several habitat units and to multiply the mean population per unit by the total number of units to arrive at a total figure for the population. It is, therefore, essential to take into consideration the efficiency of the sampling method used and the total extent of the habitat. Since no sampling method can be 100 per cent efficient, estimates of populations by random sampling will always be underestimates. The smaller the number of habitat units sampled the greater will be the exaggeration of this.

Sampling techniques are many and varied. A classification of them is possible on the basis of

    (A) The habitat to be sampled
        1. Air, soil or water.
        2. Other organism on which the sampled organism lives.
            (a) Plant
            (b) Animal
    (B) The organism to be sampled e.g. bird, insect,

A great deal of information is available in the literature which should be consulted for appropriate techniques for a particular investigation. Special attention should be given to the estimation of errors inherent in particular techniques.

3. Estimates of population from "nearest neighbour" techniques.

Either the distance between a randomly selected individual and its nearest neighbour is measured, or the distance from a point to the nearest individual, or predetermined number of individuals, is measured. Obviously it is only possible to use this approach with stationary, or slow moving organisms, or on well-marked colonies, e.g. snails, ant nests, gull nests.

Using the first method, the true "nearest neighbour" technique, density per unit area (m) may be estimated by

$$m = \frac{1}{4\,\bar{r}^2}$$

where $\bar{r}$ is the mean distance between nearest neighbours. Using the second method of selecting a random point m may be estimated by

$$m = \frac{n-1}{\pi} \times \frac{1}{a_n^2}$$

where n is the rank of the individual from the randomly selected point i.e. the nearest $= 1$, the next $= 2$, and so on, and a is the distance between the randomly selected point and the individual. To reduce calculation and to give reasonable reliability the third, fourth, or fifth nearest individuals should be used.

4. Relative estimates of populations

Relative estimates frequently provide a good return of data for the amount of time spent on them since most trapping methods used serve to concentrate animals and collect continuously. The interpretation of the data obtained depends upon

(A) changes in the actual population numbers.
(B) changes in the numbers in a particular phase or life stage.
(C) changes in activity.
(D) changes in the efficiency of trapping or searching.
(E) the response of individuals to traps i.e. some become trap-shy or trap-prone.

These potential sources of variation in results make comparison between species and between habitats on the basis of trapping data especially liable to unknown degrees of error.

Where sources of error are known, they can be corrected for and absolute estimates of populations may be derived from relative estimates. Where possible, the most reliable way of doing this is to operate direct absolute estimate methods in parallel with a relative estimate method and to draw the regression of the one on the other. The relative method may then be continued alone and absolute values derived from its results.

Yapp has devised a field operational refinement of searching by visual observation, the line transect, which gives an estimate of absolute population directly. When an observer walks in a straight line at a constant speed the number of animals seen will depend upon their average speed, his speed, the distance over which he can recognize them, and the density of the animal. The relations of these variable can be expressed as density of population, $D = \dfrac{Z}{2RV}$

where Z = number of encounters per unit time, R = the radial distance within which the animal must approach for an encounter with the observer to occur, and V = the average velocity of the animal relative to the observer, which is given by

$$V^2 = \bar{u}^2 + {}^{w2}$$

Where $\bar{u}$ = average velocity of the observer and ^{w2} = average velocity of the animal.

Line transects have been used extensively for bird populations. The speed of organism and observer should be as dissimilar as possible and this may produce difficulties in extending the method to some insects, especially if, in addition, the average speed of the animal is difficult to determine.

The final relative method from which data can be converted to absolute values which must be mentioned is removal trapping. If a known number of animals are removed from a population on each trapping occasion then the rate at which trap catches fall will be related to the size of the total population and the number removed. Although several other factors are also important the greatest limitation on removal trapping estimates in practice is the requirement that all animals must have an equal chance of being caught. Certain trapping methods e.g. sweep nets, only collect from parts of a habitat, and consequently consistently miss some individuals, while trap-shyness may be found in populations of vertebrates. The interpretation of removal trapping data must, therefore, be undertaken with care.

Summary

1. Animal ecological techniques are governed largely by movement, and size of individuals.
2. Observation of animals requires smooth, slow and quiet movement and must frequently be done at second hand through instruments.
3. Absolute estimates of animal numbers are provided by
    (A) Capture—recapture techniques. The interpretation of results is especially difficult.
    (B) Unit sampling of habitats.
    (C) "Nearest neighbour" techniques.
    (D) Relative estimates which may, in certain circumstances, be converted into absolute estimates.

Chapter VIII

PART III

Ecology and man

THE discussion of the ecology of man himself, as an animal species, must be left to another volume. It is time, however, to briefly consider why ecology, as a branch of biological science, is important to man and to assess how great is its importance.

In the first part of this book we considered many examples of ecological study. Some of these, for example the distribution of lichens and of peppered moths in areas with atmospheres polluted by industry, have quite obvious links with what effect man is having on his environment. Others—for example the study of the homostyle-heterostyle system in the primrose or the input and output of energy from a spring —are, at first sight, not relevant to human existence.

Consideration of the implications of the approaches to ecology which we have explored, and of the concepts derived from the information obtained, shows how closely connected man is with the other organisms around him. Creation of farms, plantations, buildings, cities, roads and so on alters the environment for other organisms as well as for man. The pouring of hot gases into the air, and hot water and effluent in rivers, lakes and the sea, changes the habitat on every scale, from ecosystem to micro-habitat. Because man is the final consumer in many food-chains, he is ultimately linked together with other life, to the energy derived from the sun by autotrophs. His actions cause considerable modifications of the food-web and consequently of energy flow, which remain largely hidden from his immediate view but may, nonetheless, be ultimately vital to his survival on the earth.

Methods of study developed in any one branch of ecology may prove useful in solving the problems of another. The investigation for purely scientific reasons of an ecological phenomenon today may be the key which resolves some problem of direct human relevance later. It is largely from pure science that the methods of applied ecology in agriculture, horticulture, forestry, and fisheries must derive.

The rapidly expanding human population of the world heightens the importance of ecology to man. On the one hand we must learn how to control the energy flow within ecosystems to provide enough food for hundreds of millions, while on the other we must study the ecological interrelationships of man himself. An optimist must assume that it will eventually prove possible to balance man's material and energy needs with the requirements of other organisms and the energy input from the sun.

In the short term it is clearly necessary to use what ecological knowledge we have to produce food, and to avoid further pollution of the environment. In addition we must maintain parts of the earth's surface where the influence of man is relatively unimportant in comparison with his influence in the great conurbations. In these areas urgently needed deeper study can be conducted and they may also serve for the recreation of man who evolved in such places and not in his cities. Unless we elevate to the first order of importance the ecological consequences of our decisions there will be little justification for optimism.

Appendix

An outline of statistics for ecology

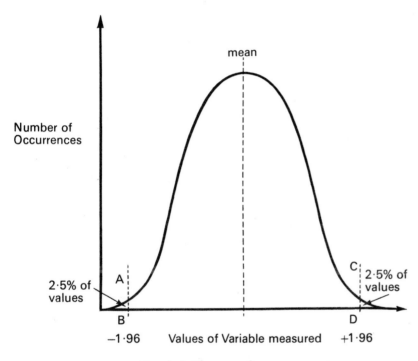

mean

Number of
Occurrences

2·5% of
values

A

C
2·5% of
values

B

D

−1·96   Values of Variable measured   +1·96

FIG. A–1. The normal curve.

Fig. A–1 shows a normal curve. Mathematically this is completely described by the mean (the average of the observations)—estimated as x̄, and the standard deviation σ, estimated as s. More than two thirds of all values fall within the range x̄ ± σ, and 95 per cent of them within

the range $\bar{x} = 1{\cdot}96\sigma$. σ (and its estimate s) therefore indicate the spread or dispersion of the observations. Assuming a population is sampled randomly then the mean and standard deviation can be estimated

$$\text{estimated mean } \bar{x} = \frac{\Sigma x}{n} \tag{1}$$

Where x represents the individual values observed, the sign Σ means "the sum of", and n is the number of observations.

Each observation differs from the estimated mean by an amount known as its deviation, $d = x - \bar{x}$. The deviation cannot be used as a measure of the dispersion of the population since by definition of the mean the sum of the deviations of all observations above and below the mean must be zero. If, however, all deviations are squared all figures become positive. The mean of the sum of the squares of the deviations, the variance, is a measure of dispersion.

$$\text{Variance } s^2 = \frac{\Sigma d^2}{n} \tag{2}$$

The variance is, however, expressed as squares of the original units which cannot be directly compared with the mean. Taking the square root of the variance restores the original units and is an estimate of the standard deviation.

$$\text{standard deviation } s = \frac{\sqrt{\Sigma d^2}}{n} \tag{3}$$

When calculating d all calculations of $x - \bar{x}$ are independent of each other except the last since this must result in $\Sigma d = 0$. In addition $\bar{x}$ is

45

an estimate of the true mean of the total population from which only a sample can be taken. The deviations from $\bar{x}$ are inevitably fractionally less than from the true mean and thus Σd^2 is slightly smaller than it should be. In compensation for this and to allow for the restriction imposed by the requirement that Σd should be zero n is replaced by n — 1 in equations (2) and (3). n — 1 is known as the degrees of freedom of the sample. Each time a further restriction is placed on the calculation n must be reduced by 1.

Probability

Probability is expressed by the symbol P. If an event is impossible then P = O. If an event is certain then P = 1. The likelihood of an event can thus be expressed on a scale varying between 0 and 1. Probability may also be expressed on a percentage scale where impossibility = 0 per cent, and certainty = 100 per cent. P = 0·95 is, therefore, the same as P = 95 per cent, or the probability is that there is a one in twenty chance of being wrong.

Fig. A–1 can be considered as a histogram, a block graph, in which the width of each block is infinitely small with the consequent impression of a smooth curve. The area of each block is proportional to the frequency, the total number, of observations at that value. It thus follows that the two halves of the curve, either side of the mean are equal in area, and the probability of any value being in either half is 0·5 or 50 per cent. The lines AB and CD enclose 95 per cent of the area below the curve and thus the probability of any value falling within this range is 0·95. The lines AB and CD are, in fact drawn 1·96 times the standard deviation from the mean. The position of other lines can be calculated for other levels of probability. It is then possible to make statements of the form

"95 per cent of all observations will be in the range of the mean ± 1·96 times the standard deviation".

The calculation of 't'

Because it is only possible to take samples from large populations, the mean and the standard deviation can only be estimated and values for P cannot be accurately realized. These effects of sample size on estimates of probability are taken into account by the calculation of 't'.

't' is in fact, influenced by the number of degrees of freedom. Table XIV has a number of horizontal lines for various degrees of freedom. The larger the sample the more nearly the value of 't' approaches that of the standard deviation until, at an infinite number of degrees of freedom, it has the same value. In general $t = \dfrac{\text{deviation from a mean}}{\text{standard deviation}}$.

Each sample taken from an infinitely large population is only one of many such possible samples each of which has its own mean. The dispersion of the means is described by the standard deviation of the means, just as the dispersion of the whole population is described by the standard deviation of the whole population.

The calculation of a value for 't' enables a level of probability to be set for the dispersion of sample means. The form of the statement of probability given above can now be adjusted to

"95 per cent of all observations will lie in the range of the mean ± t times standard deviation."

As is described in Chapter II the calculation for t can be used to estimate the level of probability of two samples being from different populations. Where samples are taken from two populations having widely differing standard deviations the analysis for t may be misleading. An advanced statistical text should be consulted in this situation.

Correlation coefficient, r

Fig. A–2 shows three situations which might result from plotting pairs of observations for the variables x and y. In Figs. A–2(a) and (b) definite trends are obvious in the scatter diagrams. In (a), as x increases y increases also; there is a positive correlation. In (b), as x increases y

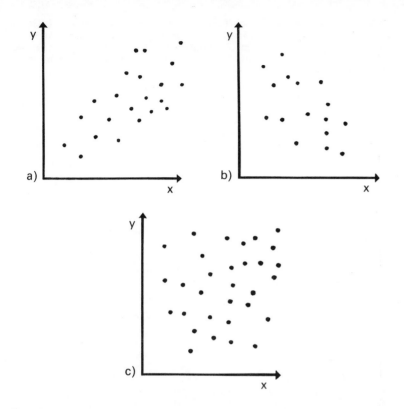

FIG. A–2. Correlation: a) positive correlation, b) negative correlation, c) does correlation exist?

decreases; there is a negative correlation. In such cases it may be sufficient to state whether the correlation is negative or positive after drawing a scatter diagram. In Fig. A–2(c) however no correlation is apparent on sight. In this case, and also where a mathematical statement of the degree of correlation is required, it is appropriate to calculate the correlation coefficient, r. This may vary from $+ 1 \cdot 0$, with perfect positive correlation, to $-1 \cdot 0$, with perfect negative correlation.

$$ r = \frac{\Sigma xy - \dfrac{(\Sigma x \times \Sigma y)}{n}}{\sqrt{\Sigma d^2 x \times \Sigma d^2 y}} $$

The only value in this equation not previously used in the calculation for t is Σxy. The significance of a calculated value of r may be estimated from Table XVI. It must be remembered that the existence of a high degree of correlation does not demonstrate a causal relation between two variables. Such relations can only be established by the application of scientific methods, backed up by the correct application of statistical techniques.

The regression equation

In many investigations it may be necessary to go one stage further and to be able to say how one factor varies as the other changes, and to calculate a value of y from a known value of x, or vice versa. This stage, a regression analysis, presupposes that one variable does in fact depend on the other.

When correlation is perfect all points will lie on a straight line. Usually, however, there is a scatter of points with general trend in a straight line. Two regression lines can be drawn through the points.
(a) The regression of y on x. This gives the line which has the minimum sums of squares of deviation of y.
(b) The regression of x on y. This gives the line which has the minimum sums of squares of deviations of x.

Regression y on x. $y = \bar{y} + b (x - \bar{x})$

The factor b gives the slope of the regression line and is calculated

$$ b = \frac{\Sigma d_x\, d_y}{\Sigma d^2_x} $$

Regression x on y. $x = \bar{x} + b (y - \bar{y})$
and b is calculated

$$ b = \frac{\Sigma d_x\, d_y}{\Sigma d^2_y} $$

When the regression line is not a straight line e.g. in the growth of bacteria, more advanced statistical techniques are required.

The Chi-squared test

This test is especially suited to the comparison of observed and expected ratios between categories of individuals.

Chi-squared is calculated

$$\chi^2 = \Sigma \frac{(\text{Observed number} - \text{Expected number})^2}{\text{Expected number}}$$

Calculations are carried out by constructing a contingency table as in Table XII. Contingency tables are described first by the number of rows and second by the number of columns. Table XII is thus a 1×5 contingency table. 2×2 contingency tables are common in biological work. In these cases the number of degrees of freedom is always 1.

TABLE XII

Moths caught of light traps at different heights above the ground, and the calculation of χ^2.

| | Light trap height in feet above ground. | | | | | |
	0	2	4	6	8	Totals
Observed (O) no. s.	24	8	25	23	20	100
Expected (E) no. s.	20	20	20	20	20	100
Deviation (O–E)	4	–12	5	3	0	0
(O–E)²	16	144	25	9	0	
(O–E)²/E	0·80	7·02	1·25	0·45	0	9·52=χ^2

With 4 degrees of freedom, using Table XV, $\chi^2 = 9\cdot52$ is between $P = 0\cdot05$ and $0\cdot02$ i.e. the initial null hypothesis that there is no difference between catches at various trap heights is untenable. The largest contribution to χ^2 comes from the trap at 2 feet above the ground and one may be confident that this is catching fewer moths than the others.

TABLE XIII

Random Numbers

16 16	57 04	81 71	17 46	53 29	73 46	42 73	77 63	62 58	60 59
98 63	89 52	77 23	61 08	63 90	80 38	42 71	85 70	04 81	05 50
01 03	09 35	02 54	51 96	92 75	58 29	24 23	25 19	89 97	91 29
29 07	16 34	49 22	52 96	89 34	17 11	06 91	24 38	55 06	83 59
72 61	80 54	70 99	24 64	11 38	83 65	27 23	40 37	84 58	48 53
71 11	41 82	79 37	00 45	98 54	52 89	26 34	40 13	60 38	08 86
61 05	66 18	76 82	11 18	61 90	90 63	78 57	32 06	39 95	75 94
81 89	42 34	00 49	97 53	33 16	26 91	57 58	42 48	51 05	48 27
10 24	90 84	22 16	26 96	54 11	01 96	58 81	37 97	80 98	72 81
14 28	33 43	01 32	58 39	19 54	56 57	23 58	24 87	77 36	20 97
35 41	17 89	87 04	28 32	13 45	59 03	91 08	69 24	84 44	42 83
07 89	36 87	98 73	77 64	75 19	05 61	11 64	31 75	49 38	96 60
27 59	15 58	19 68	95 47	25 69	11 90	26 19	07 40	83 59	90 95
95 98	45 52	27 35	86 81	16 29	37 60	39 35	05 24	49 00	29 07
12 95	72 72	81 84	36 58	05 10	70 50	31 04	12 67	74 01	72 90
35 23	06 68	52 50	39 55	92 28	28 89	64 87	80 00	84 53	97 97
86 33	95 73	80 92	26 49	54 50	41 21	06 62	73 91	35 05	21 37
02 82	96 23	16 46	15 51	60 31	55 27	84 14	71 58	94 71	48 35
44 46	34 96	32 68	48 22	40 17	43 25	33 31	26 26	59 34	99 00
08 77	07 19	94 46	17 51	03 73	99 89	28 44	16 87	56 16	56 09
61 59	37 08	08 46	56 76	29 48	33 87	70 79	03 80	96 81	79 68
67 70	18 01	67 19	29 49	58 67	08 56	27 24	20 70	46 31	04 32
23 09	08 79	18 78	00 32	86 74	78 55	55 72	58 54	76 07	53 73
89 40	26 39	74 58	59 55	87 11	74 06	49 46	31 94	86 66	66 97
84 95	66 42	90 74	13 71	00 71	24 41	67 62	38 92	39 26	30 29

```
52 14   49 02   19 31   28 15   51 01   19 09   97 94   52 43   22 21   17 66
89 56   31 41   37 87   28 16   62 48   01 84   46 06   04 39   94 10   76 21
65 94   05 93   06 68   34 72   73 17   65 34   00 65   75 78   23 97   13 04
13 08   15 75   02 83   48 26   53 77   62 96   56 52   28 26   12 15   75 53
03 18   33 57   16 71   60 27   15 18   39 32   37 01   05 86   25 14   35 41

10 04   00 95   85 04   32 80   19 01   85 03   29 29   80 04   21 52   14 76
23 94   97 28   60 43   42 25   26 48   48 13   34 68   39 22   74 85   03 25
35 63   42 90   90 74   33 17   58 77   83 36   76 22   00 89   61 55   13 17
42 86   03 36   45 33   60 77   72 92   10 76   22 55   11 00   37 60   47 73
67 26   92 87   09 96   85 37   82 61   39 01   70 05   12 66   17 39   99 34

91 93   88 56   35 76   97 35   19 37   14 66   07 57   24 41   06 90   07 72
37 14   73 35   32 01   07 94   78 28   90 33   71 56   63 77   89 24   24 28
07 46   50 58   08 73   42 97   20 42   64 68   48 35   04 38   28 28   36 94
92 18   09 46   94 95   17 41   28 60   67 94   26 54   63 70   84 73   76 61
00 49   98 43   39 67   68 40   41 31   92 28   49 57   15 55   11 81   41 89

08 59   41 41   33 59   43 28   14 51   02 71   24 45   41 57   22 11   79 79
67 05   19 54   32 33   34 68   27 93   39 35   62 51   35 55   40 99   46 19
24 99   48 06   96 41   21 25   29 03   57 71   96 49   94 74   98 90   21 52
65 86   27 46   70 93   27 39   64 37   01 63   21 03   43 78   18 74   77 07
52 70   03 20   84 96   14 37   51 05   63 99   81 02   84 56   17 78   48 45

32 88   29 93   58 21   71 05   68 58   79 08   86 37   98 76   70 45   66 23
54 16   39 40   98 57   02 05   65 15   73 23   51 51   75 06   38 13   51 68
95 22   18 59   54 57   44 22   72 35   81 24   14 94   24 04   42 26   92 14
93 10   27 94   90 45   39 33   50 26   88 46   90 57   40 47   71 63   62 59
19 20   85 20   15 67   78 03   32 23   50 59   24 83   64 99   18 00   78 50
```

Each digit is an independent sample from a population in which the digits 0 to 9 are equally likely, that is each has a probability of $\frac{1}{10}$.

After Lindley D. V. and Millar, J. C. P. *Cambridge Elementary Statistical Tables* Table 8, Cambridge University Press.

Table XIV is abridged from Table III of Fisher and Yates: *Statistical Tables for Biological, Agricultural and Medical Research*, Oliver & Boyd Ltd, Edinburgh, by permission of the authors and publishers.

Distribution of: t

TABLE XIV

Degrees of freedom	Probability, p				
	0·1	0·05	0·02	0·01	0·001
1	6·31	12·71	31·82	63·66	636·62
2	2·92	4·30	6·97	9·93	31·60
3	2·35	3·18	4·54	5·84	12·92
4	2·13	2·78	3·75	4·60	8·61
5	2·02	2·57	3·37	4·03	6·87
6	1·94	2·45	3·14	3·71	5·96
7	1·89	2·37	3·00	3·50	5·41
8	1·86	2·31	2·90	3·36	5·04
9	1·83	2·26	2·82	3·25	4·78
10	1·81	2·23	2·76	3·17	4·59
11	1·80	2·20	2·72	3·11	4·44
12	1·78	2·18	2·68	3·06	4·32
13	1·77	2·16	2·65	3·01	4·22
14	1·76	2·14	2·62	2·98	4·14
15	1·75	2·13	2·60	2·95	4·07
16	1·75	2·12	2·58	2·92	4·02
17	1·74	2·11	2·57	2·90	3·97
18	1·73	2·10	2·55	2·88	3·92
19	1·73	2·09	2·54	2·86	3·88
20	1·72	2·09	2·53	2·85	3·85
21	1·72	2·08	2·52	2·83	3·82
22	1·72	2·07	2·51	2·82	3·79
23	1·71	2·07	2·50	2·81	3·77
24	1·71	2·06	2·49	2·80	3·75
25	1·71	2·06	2·49	2·79	3·73
26	1·71	2·06	2·48	2·78	3·71
27	1·70	2·05	2·47	2·77	3·69
28	1·70	2·05	2·47	2·76	3·67
29	1·70	2·05	2·46	2·76	3·66
30	1·70	2·04	2·46	2·75	3·65
40	1·68	2·02	2·42	2·70	3·55
60	1·67	2·00	2·39	2·66	3·46
120	1·66	1·98	2·36	2·62	3·37
∞	1·65	1·96	2·33	2·58	3·29

TABLE XV

Distribution of χ^2

Degrees of freedom

Probability, p

Degrees of freedom	0·99	0·98	0·95	0·90	0·80	0·50	0·20	0·10	0·05	0·02	0·01	0·001
1	0·000	0·001	0·004	0·016	0·064	0·455	1·64	2·71	3·84	5·41	6·64	10·83
2	0·020	0·040	0·103	0·211	0·446	1·386	3·22	4·61	5·99	7·82	9·21	13·82
3	0·115	0·185	0·352	0·584	1·005	2·366	4·64	6·25	7·82	9·84	11·35	16·27
4	0·297	0·429	0·711	1·064	1·649	3·357	5·99	7·78	9·49	11·67	13·28	18·47
5	0·554	0·752	1·145	1·610	2·343	4·351	7·29	9·24	11·07	13·39	15·09	20·52
6	0·872	1·134	1·635	2·204	3·070	5·35	8·56	10·65	12·59	15·03	16·81	22·46
7	1·239	1·564	2·167	2·833	3·822	6·35	9·80	12·02	14·07	16·62	18·48	24·32
8	1·646	2·032	2·733	3·490	4·594	7·34	11·03	13·36	15·51	18·17	20·09	26·13
9	2·088	2·532	3·325	4·168	5·380	8·34	12·24	14·68	16·92	19·68	21·67	27·88
10	2·558	3·059	3·940	4·865	6·179	9·34	13·44	15·99	18·31	21·16	23·21	29·59
11	3·05	3·61	4·58	5·58	6·99	10·34	14·63	17·28	19·68	22·62	24·73	31·26
12	3·57	4·18	5·23	6·30	7·81	11·34	15·81	18·55	21·03	24·05	26·22	32·91
13	4·11	4·77	5·89	7·04	8·63	12·34	16·99	19·81	22·36	25·47	27·69	34·53
14	4·66	5·37	6·57	7·79	9·47	13·34	18·15	21·06	23·69	26·87	29·14	36·12
15	5·23	5·99	7·26	8·55	10·31	14·34	19·31	22·31	25·00	28·26	30·58	37·70
16	5·81	6·61	7·96	9·31	11·15	15·34	20·47	23·54	26·30	29·63	32·00	39·25
17	6·41	7·26	8·67	10·09	12·00	16·34	21·62	24·77	27·59	31·00	33·41	40·79
18	7·02	7·91	9·39	10·87	12·86	17·34	22·76	25·99	28·87	32·35	34·81	42·31
19	7·63	8·57	10·12	11·65	13·72	18·34	23·90	27·20	30·14	33·69	36·19	43·82
20	8·26	9·24	10·85	12·44	14·58	19·34	25·04	28·41	31·41	35·02	37·57	45·32
21	8·90	9·92	11·59	13·24	15·45	20·34	26·17	29·62	32·67	36·34	38·93	46·80
22	9·54	10·60	12·34	14·04	16·31	21·34	27·30	30·81	33·92	37·66	40·29	48·27
23	10·20	11·29	13·09	14·85	17·19	22·34	28·43	32·01	35·17	38·97	41·64	49·73
24	10·86	11·99	13·85	15·66	18·06	23·34	29·55	33·20	36·42	40·27	42·98	51·18
25	11·52	12·70	14·61	16·47	18·94	24·34	30·68	34·38	37·65	41·57	44·31	52·62
26	12·20	13·41	15·38	17·29	19·82	25·34	31·80	35·56	38·89	42·86	45·64	54·05
27	12·88	14·13	16·15	18·11	20·70	26·34	32·91	36·74	40·11	44·14	46·96	55·48
28	13·57	14·85	16·93	18·94	21·59	27·34	34·03	37·92	41·34	45·42	48·28	56·89
29	14·26	15·57	17·71	19·77	22·48	28·34	35·14	39·09	42·56	46·69	49·59	58·30
30	14·95	16·31	18·49	20·60	23·36	29·34	36·25	40·26	43·77	47·96	50·89	59·70

Table XV is abridged from Table IV of Fisher and Yates: *Statistical Tables for Biological, Agricultural and Medical Research*, Oliver & Boyd Ltd, Edinburgh, by permission of the authors and publishers.

The correlation coefficient, *r* **TABLE XVI**

Degrees of freedom	Probability, *p*				
	0·1	0·05	0·02	0·01	0·001
1	0·988	0·997	1·000	1·000	1·000
2	0·900	0·950	0·980	0·990	0·999
3	0·805	0·878	0·934	0·959	0·991
4	0·729	0·811	0·882	0·917	0·974
5	0·669	0·755	0·833	0·875	0·951
6	0·622	0·707	0·789	0·834	0·925
7	0·582	0·666	0·750	0·798	0·898
8	0·549	0·632	0·716	0·765	0·872
9	0·521	0·602	0·685	0·735	0·847
10	0·497	0·576	0·658	0·708	0·823
11	0·476	0·553	0·634	0·684	0·801
12	0·458	0·532	0·612	0·661	0·780
13	0·441	0·514	0·592	0·641	0·760
14	0·426	0·497	0·574	0·623	0·742
15	0·412	0·482	0·558	0·606	0·725
16	0·400	0·468	0·543	0·590	0·708
17	0·389	0·456	0·529	0·575	0·693
18	0·378	0·444	0·516	0·561	0·679
19	0·369	0·433	0·503	0·549	0·665
20	0·360	0·423	0·492	0·537	0·652
25	0·323	0·381	0·445	0·487	0·597
30	0·296	0·349	0·409	0·449	0·554
35	0·275	0·325	0·381	0·418	0·519
40	0·257	0·304	0·358	0·393	0·490
45	0·243	0·288	0·338	0·372	0·465
50	0·231	0·273	0·322	0·354	0·443
60	0·211	0·250	0·295	0·325	0·408
70	0·195	0·232	0·274	0·302	0·380
80	0·183	0·217	0·257	0·283	0·357
90	0·173	0·205	0·242	0·267	0·338
100	0·164	0·195	0·230	0·254	0·321

Table XVI is abridged from Table VII of Fisher and Yates: *Statistical Tables for Biological, Agricultural and Medical Research*, Oliver & Boyd Ltd, Edinburgh, by permission of the authors and publishers.

Glossary of terms

N. B. Where possible these brief definitions should be expanded by reference to the word in context.

Autecology—The study of the ecology of a single species.

Bio-geochemical cycle—The circulation of an element through living and non-living materials.

Biosphere—Region above, on, and a little below the earth's surface in which life can exist.

Biotic potential—The ability of a population to increase (k_n).

Carrying capacity—The upper limit of a population.

Climax community—More or less stable community developing at the end of a succession.

Cline—A continuous gradation of variation of an organism.

Community—A characteristic assemblage of species.

Competition—Antagonistic interaction between organisms.

Cryptic colouration—Protective colouration by camouflage.

Degrees of freedom—The total number of numerical values less the number of restrictions placed on them.

Deme—A breeding unit of a species smaller than the total population. Results from uneven distribution of a species.

Density (of a population)—The number or biomass of organisms in a unit space.

Density dependent factor—A factor which varies with the density of the population tending to alter the growth rate of the population. (The converse is a density independent factor.)

Detritus feeders (Decomposers)—Organisms feeding on the dead remains of other organisms. Includes both saprophytes and saprozoites.

Dispersal—The mode by which dispersion arises.

Dispersion (see also distribution)—The pattern of distribution of a species.

Distribution (see dispersion)—The geographical location of individuals of a species.

Ecological dominant—Species tending to control energy flow within a community.

Ecological efficiency—The energy transferred from a trophic level to the next trophic level, expressed as a percentage of the total energy lost by the lower trophic level.

Ecosystem—A complex of living and inert materials which make up a more or less balanced unit.

Ecotone—A transition zone between communities.

Ecotype—An adaptation of a plant species developed under defined ecological conditions.

Effective environment—Habitat factors of importance to an organism.

Energy flow—Movement of energy into, out of, and between trophic levels.

Environment—The sum total of influences on an organism, animate and inanimate.

Environmental resistance—The resistance to its own growth created by a growing population (k_c).

Food web—The pattern resulting from the occurence of an organism in more than one food chain.

Habitat—A unit of the environment.

Habitat factor—A definable aspect of the environment having effect on an organism.

Homeostasis, Homeostatic behaviour—The in-built tendency to maintain the status-quo. Behaviour resulting in the maintenance of the status-quo.

Industrial melanism—The evolution of dark forms of organisms, especially insects, in areas of high atmospheric pollution.

Limiting factors—Habitat factors setting limits to the growth of organisms either by their scarcity or by their excess.

Mean—The average value of the numerical observations.

Micro-habitat—A small unit of the environment.

Mimicry—The copying of protective and warning colourations.

Nett productivity—The rate of storage of energy in living tissues over a given period of time.

Niche—The ecological role of an organism separated in space or time from similar roles carried out by other organisms e.g. predation by hawks during the day and by owls at night.

Normal distribution, Normal curve—A distribution of a population of numerical values having a bell-shaped curve when drawn and completely described by its mean, and standard deviation.

Null hypothesis—An hypothesis that there is no significant difference between two sets of data.

Point frame—A frame on which are mounted one or more optical cross-wires, or pins, for vegetation sampling.

Polymorphism—The occurrence together in the same habitat, at the same time, of two or more distinct forms of a species in such proportions that the rarest of them cannot be maintained by recurrent mutation of genes.

Population—(a) a group of organisms of the same species occupying a particular space.
(b) statistically, the theoretical possible total number individual values of any quantity measured.

Primary consumer—An organism deriving its energy directly from primary producers. Herbivore.

Primary producer, producer—An organism converting radiant energy from the sun directly into chemical energy. An Autotroph.

Primary succession—Succession initiated on an area not previously occupied by organisms.

Pyramid of biomass—The decreasing absolute weight of living organism within the sequence producer, primary consumer, higher level consumers.

Pyramid of energy—The decreasing energy content of the standing crop of organisms at each trophic level in the sequence producer, primary consumer, higher level consumers.

Pyramid of numbers—The decrease in absolute numbers of organisms at each level through the sequence producer, primary consumer, higher level consumers.

Quadrat—A defined area for vegetation sampling, usually square in shape.

Random sampling—Measurement of some of a population of organisms or values in such a way that the taking of one measurement has no influence on the taking of any other.

Regular sampling—Measurement of some of a population of organisms or values in such a way that the taking of any one measurement is entirely dependent on the taking of the others.

Secondary succession—Succession initiated on an area already occupied by organisms.

Sere—A succession of plant communities in a well-defined habitat e.g. a psammosere on sand dunes.

Standard deviation—Square root of the sum of the squares of the deviations from the mean divided by the total number of observations.

Standing crop—The total energy within a trophic level at any given moment of time.

Stratified random sampling—Random sampling within subjectively defined sections of the area under investigation.

Succession—A sequence of organisms occupying a habitat as a consequence of their predecessor organisms altering that habitat in their favour.

Synecology—Study of the ecological interrelationships of all organisms in an area.

Trophic level—Feeding type and position within a food web e.g. secondary carnivorous consumer.

Vagility—The potential of an individual organism for movement.

Variance—Sum of the squares of the deviations from the mean divided by the total number of observations.

Bibliography of selected references

Bennett D. P., Humphries D. A. *Introduction to field biology*. Edward Arnold, 1965.

Bishop O. N. *Statistics for biology*. Longmans, 1966.

Browning T. O. *Animal Populations*. Hutchinson, 1963. Quantitative study of animal populations with considerable reference to environmental influences.

Dowdeswell W. H. *Animal Ecology* 2nd. ed. Methuen, 1961.

Elton C. *Animal Ecology*. Science Paperback Edition, 1966. A classic account of basic observational ecology.

Ford E. B. *Ecological Genetics*. Methuen, 2nd. ed. 1965. An advanced text for reference.

Kershaw K. A. *Quantitative and Dynamic Ecology*. Edward Arnold, 1964. Quantitative plant ecology.

Lewis T. and Taylor L. R. *Introduction to Experimental Ecology*. Academic Press, 1967. Useful for reference to statistical techniques of analysis and suggested exercises.

Macfadyen A. *Animal Ecology: Aims and methods*. Pitman, 2nd. ed. 1963. An advanced general survey.

Odum E. P. *Fundamentals of Ecology*. W. B. Saunders, 2nd. ed. 1959. Broad survey based upon the energetic approach.

Odum E. P. *Ecology*. Holt, Reinhart and Winston, 1963. An introductory text based upon the energetic approach.

Parr, M. J., Gaskell, T. J., George, B. J. *Capture-Recapture methods of estimating animal numbers*, J. Biol. Educ.1968 *2* pp. 95–117.

Phillipson J. *Ecological Energetics*. Edward Arnold, 1966. A brief survey of the energetic approach.

Sheppard P. M. *Natural Selection and Heredity*. Hutchinson, 3rd. ed. 1967. Surveys the background to studies in evolution and ecological genetics.

Southward T. R. E. *Ecological Methods: with particular reference to the study of insect populations*.

The publications of the British Ecological Society, and the Institute of Biology are especially useful for reference.

Journal of Ecology.

Journal of Animal Ecology.

Journal of Applied Ecology.

Symposia of the British Ecological Society.

All published by Blackwell Scientific Publications for the British Ecological Society.

Journal of Biological Education.

Published by Academic Press for the Institute of Biology.

Index

Printed in Great Britain by A. Wheaton & Co., Exeter.